KT-237-303

EXPEDITION GUIDE
by Wally Keay

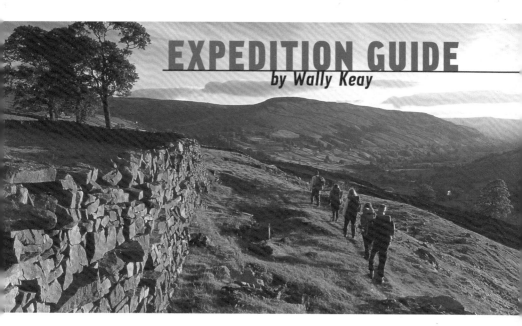

The Duke of Edinburgh's Award,
Gulliver House, Madeira Walk, Windsor,
Berkshire SL4 1EU

Cover Photograph: Joe Cornish

Printed and bound in Great Britain
by Sterling Press, Wellingborough,
Northamptonshire.

Fourth Edition February 2000
© Wally Keay 2000
© The Duke of Edinburgh's Award 2000

Fourth Edition ISBN 0 905425 14 6
(First Edition ISBN 0 905425 01 4
Third Edition ISBN 0 319 00896 7)

203 592

INTRODUCTION

This *Expedition Guide*, like its predecessor, is directed towards the participants themselves, placing information and skills in their hands so that they may take possession of their ventures and be less dependant on leader intervention.

This 4th edition incorporates all the recent changes in the conditions which take the Award into the next millennium, including the opportunity to become involved in Exploration at Bronze level. It still retains the emphasis on the philosophy and educational aims of the Expeditions Section and the value of utilising all the different options including the different modes of travel.

Most of the advice on the training and skills remains the same, as does the emphasis on the supporting roles and responsibilities of the Instructor, Supervisor and Assessor. These chapters are a recognition of the continuous hard work of these dedicated volunteers, without whom the Award could not operate.

There is a wealth of competence and experience amongst the Instructors in the Award so there will always be wide divergences of opinion in the approach to technique and good practice. The intention of this *Expedition Guide* is to provide basic information for participants in the early stages of their involvement with the Award until they are able to rely on their own skills and experience. Where the *Guide* assists this process I thank all who have helped, where there are shortcomings, as the author, I accept responsibility.

The *Expedition Guide* will have fulfilled its purpose if it adds to the enjoyment and enhances the quality of the experience within the Expeditions Section, and all the effort will have been worthwhile.

Wally Keay

Every participant in the Award Scheme has to take part in some form of Expedition. This has provided the opportunity for many young people to find their way through unfamiliar country and to fulfil their spirit of adventure and discovery. It has encouraged over two and a half million participants to overcome challenges and to discover their own hidden strengths and talents.

This new and expanded Expedition Guide sets out to explain the risks and challenges in undertaking expeditions on both land and water. It is intended to help Award participants and others, and all involved in expedition training, to prepare and plan demanding ventures, which they can then undertake with confidence.

Safety is the absolute priority in all expedition work. The advice in this Guide is the product of forty years' experience, and careful observance should help to maintain the very satisfactory record of safety in Award expeditions.

The knowledge and skills acquired from a study of this Guide will do much to ensure that each expedition is a pleasure and an enlightening experience and it might, perhaps, even encourage a desire to continue similar forms of adventurous activity in later life.

Contents

Part **4** - ADVICE AND SKILLS ASSOCIATED WITH THE MODE OF TRAVEL

Land Based Ventures

Water Based Ventures

Part **5** - YOUR VENTURE - PREPARATION AND PLANNING

Part **6** - SUPPORTING ROLES AND RESPONSIBILITIES

Part **7** - APPENDICES

'The essence of Berghaus is encapsulated in the words 'OUT THERE'. The company was founded in Newcastle-upon-Tyne back in 1966 and since then has designed and marketed products for a wide range of outdoor activities. What really inspires the people who work at Berghaus is news of those people who venture into the outdoors and make the most of that environment. To see our products being used on the hills and mountains is extremely fulfilling and brings a real meaning to what we do.

'The Duke of Edinburgh Award provides young people with the ideal opportunity to learn about the outdoors and then get out there themselves and experience the enjoyment of hiking, camping and other related activities. Berghaus is delighted to be able to support the scheme and help The Duke of Edinburgh's Award in its ongoing campaign to support young people as they explore the world around them.'

Peter Smith, Brand President, Berghaus

My life has been filled with a love of adventure and climbing, of going to places where no one has ever been before, of constantly pushing the limits. It is not a matter of doing something dangerous, dangerously. It is about taking carefully calculated risks and optimising chances of success.

'The Duke of Edinburgh Award enjoys an excellent national reputation and offers an ideal grounding for young people in the principals of outdoor activity. The Award not only succeeds in getting people into the great outdoors, but, just as importantly, also furnishes them with the skills that will allow them to make the most of such an environment, encouraging them to be independent. In addition, The Duke of Edinburgh's Award teaches young people to respect the environment that they are exploring and to make the most of it without spoiling it for others.'

Sir Chris Bonington, Chairman, Berghaus

GET LAYERED

To really get the most out of the outdoors, it's vital to dress correctly. The Berghaus Layering System recommends three crucial layers to provide essential comfort, warmth and protection.

COMFORT LAYER

The comfort layer should be made from a performance synthetic fabric such as ACL1000 that wicks perspiration away from your skin and out through the next layer of clothing. By keeping your body dry you'll also keep it comfortable. Berghaus Active Comfort Layer (ACL) tops and pants are perfect for providing next-to-the-skin comfort.

WARMTH LAYER

Warmth layer garments should be lightweight, warm and dry quickly helping you to keep in body heat. They should also let perspiration vapour escape from the comfort layer underneath. Berghaus fleece jackets made from Polartec® are ideal warmth layer garments.

PROTECTION LAYER

The waterproof and windproof outer layer is designed to provide protection from the elements. It should also be highly breathable letting perspiration vapour escape easily from lower layers. Perfect examples are Berghaus Aquafoil® and Gore-Tex® jackets and trousers.

w w w . b e r g h a u s . c o m

ACKNOWLEDGEMENTS

A book of this nature is the product of many willing people who have given generously of their time and expertise.

Without the help of Richard White and Penny Warman this book could not have been written. Richard's understanding of publishing and his experience in bringing together all aspects of the book was essential. Penny's attention to detail and her patient support was invaluable. Their expertise and dedication is evident in all aspects of the Guide.

I enjoy trying to express skills and concepts in a visual form. I have been helped enormously with this by a number of people who have translated my drawings and sketches. My thanks go to Brian King for the illustrations, Nicholas Gair for the technical drawings and Tony and Simon Winnall for the computer generated diagrams.

The advice of the Expedition Advisory Panel and Technical & Safety Committee is gratefully acknowledged, in particular John Driscoll, Stuart Briggs and Alan Surtees. My grateful thanks are also due to the Award Secretaries and Regional Officers, especially Andy Reade, Sandra Skinner, Shirley Price, Janet Shepherd, Dudley Hewitt and Paul Redrup. In addition, there are many competent people throughout the Award who have made a valuable contribution.

The attractive presentation of this book has been the responsibility of Dennis Sheppard of T-Square Design and Dave Wood, the Award Communications Officer. My thanks are also due to Maralyn Lewis, Joanna Buckley and Steve Sharp for commenting on the text and undertaking the patient and often tedious task of proof-reading.

I am grateful to George Fisher of Keswick, Alan Day, Tony Wale and Silva (UK) Ltd. for their advice and loan of equipment. Special thanks are due to Berghaus for their generous financial support in the production of this book.

The following have contributed to the photographs which have enhanced the quality of the book: D.M. Boyes, Joe Cornish, Margaret Dixon, Deb Dowdall, John Driscoll, D.W. Elson, William Fediw, Alistair Lee, Shirley McKiernon, Roger Street and Maggie Willis. **The Fuji/Halina Bursary photographers:** Tony French, Jennie Hills and Paul Smith; **The National Trust Photo Library:** Andrew Butler, Neil Campbell-Sharp, Robert Eames, Jim Hallett, Nick Meers, David Norton, Ian Shaw and Ian West; **Oxford Scientific Films:** David Boag, Kenneth Day and Richard Packwood; **Still Moving Picture Company:** Doug Corrance, Glyn Satterley; **The Railway Picture Library; North York Moors National Park; Yorkshire Dales National Park; Silva (UK) Ltd.; The Gateway Award:** Alan Lewis, **The Sea Cadets** and **Berghaus**.

Finally I would like to thank my son John for keeping my word-processor functioning, my daughter Jane for assisting with my research and my sister Moira Gresswell for maintaining me in an operational condition.

Wally Keay 2000

EXPEDITION GUIDE

THE DUKE OF EDINBURGH'S AWARD

Part

1

Joe Cornish / National Trust

THE

AWARD

SCHEME

EXPEDITIONS

SECTION

Part 1

The Award Scheme Expeditions Section

Aims and Objectives

The aim of the Expeditions Section is:

To encourage a spirit of adventure and discovery

THE PRINCIPLES

All ventures involve self-reliant journeying in the countryside, conceived with a purpose and undertaken on foot, by cycle, on horseback or on water by the participants' own physical efforts and without motorised assistance. The venture must present the participants with an appropriate challenge in terms of purpose and achievement with the minimum of Award Leader intervention. All participants must produce a presentation of their venture.

The venture demands:

- Enterprise and imagination in concept.
- Forethought, careful attention to detail and organisational ability in preparation.
- Preparatory training, both theoretical and practical, leading to the ability to journey safely in the chosen environment.
- Shared responsibility for the venture, leadership from within the group, self-reliance and co-operation amongst those taking part.
- Determination in execution.
- A reflective report related to the purpose of the venture.

The aim and the objectives have remained the same since the Award Schemes' inception which is remarkable tribute to the wisdom of those who initially drafted the challenge of the Expeditions Section.

BENEFITS TO YOUNG PEOPLE

Although the challenges are expressed in terms of physical demands by travelling for a given distance or number of hours, the Award is equally concerned with the development of the individual, teamwork and the social interaction of the group.

The Expeditions Section should provide opportunities to:

- **demonstrate enterprise** – the wide choice of ventures provide the opportunity for young people to demonstrate the greatest possible enterprise and imagination,
- **work as a member of the team** – all ventures are a group effort and all must work together in a team to ensure a successful outcome,
- **respond to a challenge** – the initial level of challenge is determined by the group but the weather and the demanding surroundings always necessitate the group responding to unforeseen challenges,
- **develop self-reliance** – a progressive programme of training with diminishing external Leader involvement enables the young people to become more self-reliant until the group is able to carry out the qualifying venture unaccompanied,
- **develop leadership skills** – members of the group have opportunities to take a leading role during different aspects of the venture so that everyone develops their own style of leadership and the ability to communicate their intentions to others,
- **recognise the needs and strengths of others** – the whole group must work together to complete the venture, so it is essential that each member recognises the strengths and weaknesses of others and are all involved in mutual support,
- **make decisions and accept the consequences** – outdoor ventures demand total commitment and groups have to learn to make decisions affecting their well-being and accept the consequences of their actions,
- **plan and execute a task** – each group has to plan and execute a venture with a clearly defined purpose. This follows a training programme from which they acquire the necessary experience so that they can complete the task by their own effort,
- **reflect on personal performance** – to increase the effectiveness of the training and the memorable experiences of their venture, the participants should have opportunities to reflect on personal performance and review their progress as a team,

- **enjoy and appreciate the countryside** – the Expeditions Section encourages positive responses to the natural world and growing environmental awareness.

ADVENTURE AND DISCOVERY

The aim of the Expeditions Section is to encourage a spirit of adventure and discovery but the challenge is expressed in terms of practical physical endeavour, a specific distance or a number of hours of travelling. The form of the challenge must never disguise the underlying aims and objectives of the Section which is concerned with adventure and discovery. Neither must there be a misunderstanding about the nature of the discovery. Discovery is not limited to finding out about the natural world; it is as much concerned with finding out about oneself, the feelings of elation and satisfaction which come from overcoming a challenge, and the self-esteem which comes from achieving a task which was not considered possible in the face adversity and danger. All the journals, logbooks and writings of explorers who have sought adventure on land, water or in the sky reveal that self-discovery and self-awareness is central to adventure and discovery.

Discovery does not have to be original, whether it is environmental, geographical or concerned with self-awareness or the understanding of others. That many others have trod the same path, or have made the same discovery, is totally irrelevant, as the vital factor is that the discovery or the experience is new to the individual.

ALL VENTURES WITHIN THE EXPEDITIONS SECTION TAKE THE FORM OF A JOURNEY

Regardless of the mode of travel, or whether it is an Expedition, Exploration or an Other Adventurous Project, all ventures take the form of journey. It is journeying which forms the fundamental concept of the Expeditions Section. All benefits, whether in personal development or in team building, which occur through the growth of interpersonal relationships, are brought about by travelling together.

Journeys provide the opportunity for young people to separate themselves from the problems and stresses of their everyday life, it is a chance for them to devote their attention to the simple but vital tasks of ensuring their own survival and that of their companions, and achieving the purpose of their venture. It is this contrast between the essential simplicity of the Expedition lifestyle and the complexity of modern life, which is one of the greatest rewards for young people.

Participants in the Award Scheme must travel by their own physical effort and the speed of travel is, of necessity, slow. This slowness is of considerable significance as it enhances the feeling of remoteness in time and distance. The benefits of the journey are inversely proportional to the speed of travel, extending the separation between the monotony and frequent boredom of the daily routine and the excitement and stimulation of a self-reliant venture. Remoteness from immediate help necessitates total commitment and, where real decisions have to be made, the consequences of those decisions have to be accepted. The requirement that accommodation will be by camping will further emphasise this self-sufficiency; to stay in a hostel would totally destroy any feelings of independence and isolation. Similarly, the intervention of adults must be limited to the minimum necessary to ensure safety and that the conditions of the Award are fulfilled. **All ventures belong to the participants.**

WORKING TOGETHER

All the Sections of the Award are designed to pose different opportunities and challenges to the individual. A significant difference between the Expeditions Section and the other Sections is that success can only be achieved by the collective effort and co-operation of a number of participants. Teamwork, so essential for the completion of the venture, must not be left to chance. The strength of the personal relationships which exist amongst the members of the group are just as essential for ensuring the success of a venture as the skills related to the training syllabus.

Because ventures demand teamwork, the Expeditions Section, by its very nature, provides the basis for working together as a team. Wherever individuals are dependent on each other for their safety and well-being, there are opportunities for team building. There are no more effective situations for establishing mutual dependence than paddling the same canoe, living together in two or three square metres of tent, righting the capsized dinghy or being at opposite ends of a life-line. The interdependence established during the training sessions and the consolidation achieved during the practice journeys is the real basis for team building.

Adventure and discovery always involve some measure of risk, no matter how small, and only the interpersonal relationships which have been forged in real situations in the frequently harsh conditions of the outdoors, can produce the trust between the members of a group which will stand up to the stresses of adversity or threatening situations. Contrived exercises in team building have no lasting substance for where there are no penalties for failure, there are no rewards for success. However, team building exercises may play a significant part in bringing a group together at the start of the training process.

In working together as a team we learn to recognise the strengths and weaknesses of our companions at the same time as we learn about ourselves. While there is an expectation that each member of a venture group will be adequately trained and competent, it is inevitable that some members will be more proficient than others in some aspects of the training syllabus. It is comparatively easy to assess the skills of navigation, or first aid and resuscitation - though the assessment may be quite different when operating under the stress of real threatening situations - it is less easy to evaluate the interpersonal skills.

Participants learn to modify their own personal preferences to meet the common aspirations of the whole group, so that all can be involved in the planning and execution of the venture. Individual members of the group make their own contribution to the collective competence of the team; frequently the success of the venture may be more dependent on the cheerfulness and the strength of purpose of a particular individual than on the obvious skills of another member of the team.

During the venture, leadership must come from within the group. The roles of leadership play such an important part in young people working together that the following chapter has been devoted solely to this aspect.

NATURE AND THE ENVIRONMENT

The aim and objectives place considerable emphasis on the enjoyment and appreciation of the countryside. Nature has always been the inspiration of artist, poet and composer, and the driving force of scientists in their endeavours to provide explanations of the natural world and its phenomena. Explorations increase the time available to study and to respond to the natural world but, whatever the form of venture or mode of travel there will, in addition, be a growing environmental awareness and, it is hoped, a personal concern for the protection and care of the countryside and the waters which surround it.

PRESENTATIONS

No venture is complete without a presentation, whatever its form. This, again, is true of the great tradition of discovery, exploration and expeditioning. Our literature continues to be enriched by the exploits of present day sailors, climbers and travellers and the accent on self-awareness, self-discovery and the insight into their companions never ceases to fascinate. Award participants may be more modest in their aspirations but their presentations are no less important.

Some reports will be based on studies and explorations and built on the observations and recordings which take place in the field; others may be photographic or visual presentations. Many are journals or accounts of the venture and may use any of a number of forms of presentation. It is important that presentations are concerned with attitudes and feelings rather than the mundane detail of process and progress. The presentation is an opportunity to review the venture and its memorable experiences and to reflect on one's personal achievements and performance, as well as that of the group.

The Expeditions Section is progressive with its three levels of Bronze, Silver and Gold. Each level of Award is more demanding than the previous level and requires a longer and more sustained effort on the part of the participants. The process of reflection and review of training, the performance of the group and personal relationships within the group, form an important means of evaluating performance. This can be used as a base for progression to the next level of Award, or in the subsequent ventures which many young people will become involved with after achieving their Gold Award.

There is often a heightened sense of awareness and an increased sensitivity of perception during adventures, and the sensory arousal which occurs is a powerful factor in enriching experiences and making them memorable. Much of the experience will be of an aesthetic nature and will be beyond our powers of expression; no recordings or observations will do justice to the view from a tent of the sun setting on a mountain landscape, of a starry night with the heavens unobscured by the scattered light and smog of our cities, or the dawn chorus in some remote valley. This not a matter for concern; those things which are most meaningful and worthwhile in our lives aren't easily measured or expressed. It is the experience itself which is essential!

The final venture within the Award should never be regarded as the end; it is just the beginning, the stepping stone to other longer and more adventurous journeys of discovery, where all the skills, training and experience gained in completing the Award become the starting point for future adventures and discovery in later life.

Joe Cornish

Leadership and Teamwork

Amongst the objectives set out in the Expeditions Section are that participants:

- Work as a member of a team.
- Develop leadership skills.
- Recognise the needs and strengths of others.
- Make decisions and accept the consequences.

All of these are inextricably linked with leadership in the same way that leadership is linked to teamwork. A successful group requires that leadership and teamwork act in balance as they are interdependent if the group is to achieve its goals and individuals make their greatest contribution. The *Award Handbook* states:

'**Develop leadership skills** - members of the group have the opportunity to exercise the role of leader during the different aspects of the venture so that everyone has a chance to develop their own style of leadership and the ability to communicate their intentions to others.'

This chapter is primarily concerned with the development of leadership skills amongst the young people involved within the Award Scheme, but the same principles and considerations apply to all leadership, and the chapter has the same relevance for Leaders and Instructors of Award Units who deliver the Scheme as it does to the participants.

There are many who believe that leaders are born not made. Certainly some aspects of personality or character facilitate taking on the mantle of leadership and make the role much easier. In everyday life, however, most people have the role of leader thrust upon them - so-called 'situational leadership' - in particular professions, by becoming school teachers, by being involved in the Award Scheme, or through promotion in the office, the factory or in sport. The majority of people who have the role of leader thrust upon them rise to the occasion and learn the role as they go along - learning on the job! It is the opportunity to practise the role of leader that is the most important factor. We may not all be born leaders, but it is possible for everyone to improve their qualities of leadership.

Adult Leaders and Instructors must ensure that all participants have an opportunity to develop their leadership skills and that the role is not confined to the most assertive or dominant. Instructors can assist in the process by delegating the role of leader to individuals during training and practice journeys, and expect similar procedures to be followed on qualifying ventures.

If the skills of leadership are to develop and not just remain a pious hope, then at least two things are necessary:

- The participants must have realistic opportunities to practise the role of leader.
- The process of leadership must be reviewed by the participants.

To enable the process of reviewing to take place, it is necessary to have some background against which discussion and evaluation can take place. Many hundreds of books and papers have been written on leadership, almost as many as on management, so it is essential, as with all things within the Expeditions Section, that everything is kept as simple as possible. There are, however, two very simple models concerned with leadership which will help to facilitate the process. One is a continuum produced by Tannenbaum and Schmidt in 1968, which deals with the style of leadership. At one end is a totally authoritarian style of leadership while at the other is a collective process of decision-making.

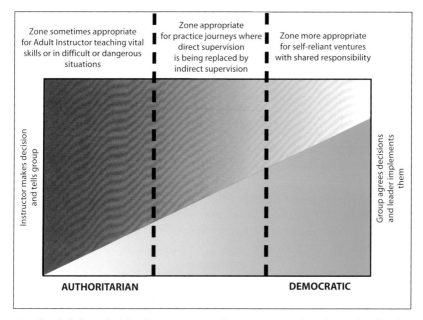

On the left-hand side there is an authoritarian style where the leader makes the decisions and orders, or commands, the action which is to be followed. At the right-hand side leadership involves the whole group being actively involved in the decision-making process and the resulting action arises from collective agreement.

Those involved in the Award Scheme, both young and old, are volunteers and it is obvious that the style of leadership towards the right-hand side of the continuum, where all are involved in the decision-making, is more appropriate for the Scheme. This applies not only to leadership between members of venture groups, but also between the Adult Leaders and the participants. It facilitates personal development, confidence and team building. It is very important that the adults in the Award intervene as little as possible in the decision-making process and leave as much as possible in the hands of the participants.

While a democratic form of leadership where the decisions are reached by agreement within the group are more suitable for the Expeditions Section, it must not be thought that an authoritarian leadership is wrong. There are times when faced with hazard or life-threatening situations such as a flooded river, a canoe pinned under a branch or an electrical storm on high exposed ground, where authoritarian leadership is the most appropriate - it may be the only style of leadership which will bring about a safe conclusion. It also has the advantage of facilitating rapid decisions.

The continuum allows any number of styles of leadership between the two extremes to operate and it is up to the leaders to make use of the style which is most appropriate for the task in hand, and to their own individual style and personality. When members of a group are reviewing individual leadership performances, they will wish to consider if the styles of leadership were appropriate to the situations.

There is another leadership model, of unknown origin, published and advocated by John Adair. It has the same essential simplicity as the first model and is just as easy to understand.

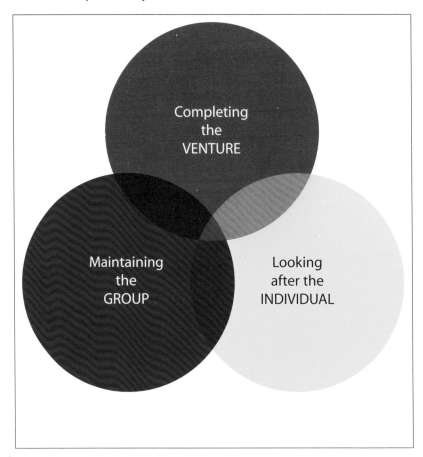

Here the role of the leader is divided into to three distinct elements:

- Achieving the objectives of the task.
- Maintaining the group.
- Satisfying individual needs.

This is a most appropriate model of leadership for the Expeditions Section. The three circles overlap but are not coincidental, indicating that the needs of the three circles are different and frequently conflict with each other.

Achieving the Objectives of the Task

The completion of a successful qualifying venture should be a clear and unambiguous task since it has been constructed around a common purpose and a journey which has been devised and agreed by all the participants within the group. Similar tasks, though possibly on a lesser scale, have to be fulfilled during the whole of the training programme and the progression to the qualifying venture. The completion of a night navigation exercise, erecting a tent in five minutes and achieving the specific aims of a practice journey all present appropriate tasks and challenges to the leader and success can be reviewed and evaluated.

Maintaining the Group

Ventures within the Expeditions Section necessitate the formation of a group as the requirements and conditions cannot be fulfilled by one person. Many of the problems which arise during ventures are rooted in the group's inability to work together; far more so than through a lack of technical competence, other than navigation. Maintaining the group as a cohesive unit with a high level of motivation is a vital aspect in the completion of a successful venture. It is the role of the leader to ensure that, when faced with adversity, the group, by its combined effort, can work together to overcome the difficulties. Similarly, when an individual has a problem, the collective effort of the group can be mobilised to help resolve the individual's problem. All Assessors and Supervisors can quote instances where the combined efforts of a group have supported individuals in difficulty to ensure a successful outcome to the venture.

The leader also has the task of ensuring that the strengths of the group are fully mobilised to achieve the common task. Teamwork is a corner-stone of the Expeditions Section and the building of a team from the first coming together is inseparable from maintaining the group while it is achieving its task. Developing good personal relationships within the group is an essential part of the process and requires a period of time if groups are to develop their own identity and be dynamic.

Satisfying Individual Needs

The needs of each member of the team have to be supported in the same way that the group has to be maintained, and it is necessary for the leader to recognise the needs and strengths of individuals. There may be a conflict between the group's need and the needs of a particular member of the group, and these conflicts need to be resolved. Individuals' needs frequently have to be secondary to those of the group and related to the completion of the task. Individuals frequently require the support and understanding of the group if the venture is to be successful and it is the leader's task to identify the problems of the individual before they jeopardise the common goal. Similarly, the leader must recognise the abilities and strengths of individuals so that they may be utilised to increase the total effectiveness of the group and enable all individuals to contribute according to their abilities.

The leader has to maintain a balance between these three elements of the leader's role if success is to be achieved, especially in the face of adversity. It is frequently both a juggling and balancing act but, out of this interplay, a development of personal attitudes and a maturity of behaviour should result and enhance all the individuals within the group.

The ability to keep these aspects in equilibrium may be a measure of good leadership. Individual members of a group should be able to assess the effectiveness of the leadership. If this leadership model is used, along with the previous one, it will enable participants to carry out a meaningful review and evaluation of each other's qualities of leadership against simple criteria which are easy to comprehend.

The process of reviewing will do more to improve the quality of leadership of individual group members than any other action, apart from the opportunity to adopt the role of leader. The learning process and the acquisition of skills, confidence and effectiveness will be speeded up immeasurably by discussion and evaluation. Leadership will no longer be an intangible quality with which some are endowed, but a skill which can be acquired and discussed in practical terms. See Chapter 6.1 - Training.

There are certain other attributes and qualities which may be developed to increase the effectiveness of leadership. These are particularly important and all young people within the Award Scheme who aspire to develop their leadership skills may find them relevant.

TECHNICAL COMPETENCE

Technical competence is a good foundation on which to build leadership. It inspires confidence, and those who are being led have a right to expect that leaders will be good at their job. Competence in navigation, what to do in an emergency, or in first aid all give rise to trust. Leaders who have ability in technical skills and where the exercise of the skills is automatic, requiring little conscious thought, are able to devote more of their attention to the needs of the group and the individuals within the group. Technical competence encourages confidence in leaders, resulting in a more assured style of leadership.

EXPERIENCE

Technical competence is not sufficient by itself; it must be enhanced by practical experience built up gradually over a period of time. There is no such thing as instant experience! Young people should avail themselves of every opportunity to extend their experience. It is important that it is developed in a variety of situations and under different conditions for, all too often, some leaders' experience consists of repeating the same exercise over and over again. A variety of experience in progressively more demanding situations enables the right skills to be applied to reach the correct solutions.

A lack of experience cannot be hidden for very long and there is nothing more demoralising than for leaders to find that members of the group have much greater experience than they have.

Where opportunities exist to observe competent leaders at work, their techniques should be studied and analysed as it is one of the most effective ways of improving ability.

PLANNING AND PREPARATION

Good leaders are distinguished by the quality of their planning and the thoroughness of their preparation. Nothing will do more to increase the confidence of the leader, and of those being led, than effective planning and preparation. It ensures that the venture runs smoothly and to plan, enhances the margins of safety and frequently turns a modest leader into a very safe and effective one.

Good preparation is time-consuming and usually does not take place sufficiently far in advance of the event, as all Wild Country Panel Secretaries will testify. The quality can be improved and the process made less time-consuming by keeping records of previous activity. Checklists of equipment are needed, never rely on memory for one usually ends up by taking ninety-nine out of the hundred and one items you should have taken and the two missing ones are vital to the well-being of an individual or the group. Contacts, addresses, telephone numbers, timings, costings and accounts of previous events form the basis on which to plan future ventures. Routines make the task less arduous.

Within the Award Scheme it is important that the participants have the responsibility for planning and preparing their ventures, whether they are practice journeys or qualifying Expeditions. Their preparation should extend to all aspects of the venture and include camp site and travel bookings as well as transport arrangements. It is this planning and preparation which is so vital to the learning process; all too often the young people are denied the opportunity to participate in the process. This said, the prudent Instructor or Supervisor will monitor the preparation carefully.

COMMUNICATION

Referring back to the beginning of the chapter, it will be noticed that the leadership aim states: 'and the ability to communicate their intentions to others'. The ability to communicate clearly and precisely is one of the hallmarks of a good leader.

All the previous abilities may be impaired if leaders cannot express their intentions clearly. Some are able to communicate better than others but, as with leadership itself, everyone can improve. Do not be discouraged if you find it difficult at first. Managers, teachers and youth leaders all experience some apprehension but it quickly disappears with practice. The observation of a few simple rules enables one to become a more effective communicator.

Paul Smith (Halina/Fuji Bursary)

Do not speak until you have thought out what you are going to say. Always have a piece of card or a notebook and pencil to hand so that you can jot down a number of items - there is nothing more irritating than the activity being interrupted continually because the leader keeps forgetting important items. It is helpful if communication can take place at a natural break in the proceedings, stopping to check the map, donning wet weather clothing etc.

Express yourself as clearly as you can as serious accidents have occurred because instructions have not been clearly expressed or understood. Communication is a two-way process and individuals should be given every opportunity to ask questions to ensure that they understand what is required of them.

There are obvious opportunities for communicating which should always be taken. The leaders should always brief the rest of the group, even if they are only fulfilling the role of leader for a couple of hours, so that each member of the group has a clear understanding of the immediate objectives whether it is a practice journey or a qualifying venture.

Briefings should always take place before an activity and if, as in the case of a practice journey or qualifying venture, the activity takes place over a number of days, there should be a briefing before the start of each day. It should include the aims and objectives and an outline of the day's or the session's programme and timings.

When communication has to take place en route or out-of-doors during training sessions, choose the most suitable place. Use a natural break where possible, and find a sheltered place out of the wind or rain where everyone can be seen and heard. Do not attempt to communicate until you have everyone's attention. Keep instructions or requests short and to the point and, above all, don't indulge in monologues.

AWARENESS..

..Of the Group

Awareness is another factor in the make-up of a good leader. and again we return to the aims of the Expeditions Section in recognising the needs and strengths of others. Awareness is principally based on observation, but hearing and listening also play their part. A developed awareness enables the leader to observe closely the needs of the group as a whole, as well as the particular needs of individuals. The person who is lagging behind, possibly showing the earliest signs of incipient hypothermia, will set alarm bells ringing as other members of the group may also be approaching a similar condition and urgent remedial action will be needed.

..Of the Individual

Sometimes the needs of an individual will conflict with the needs of the rest of the party and balancing these conflicting needs may require all the leader's skill. It is inevitable that some individuals will need more attention and care than others. Those who have medical or behavioural problems, or any other special needs, will always require careful attention.

..Of the Environment

A swollen stream, a squall in an exposed estuary or cold driving rain all pose obvious threats to a group. Leaders must be aware of what is going on around them, anticipate problems and take the necessary avoiding action. Not all dangers to a party are so dramatic. The greatest dangers nearly always arise from everyday, mundane activities such as getting in and out of buses, from walking and cycling on narrow twisting country roads and in fuelling cooking stoves. Constant alertness and the stamina never to be lulled into a false sense of security by familiar surroundings and routine practices are vital qualities in the making of a leader.

Awareness is assisted by leaders positioning themselves in the right place; some seem to be able to do this instinctively. The ability to take up a position where a hazard may arise gives an added protection to the group. Anticipation is a valuable asset and the foresight to be able to perceive threatening situations in advance all make for safer activity.

JUDGEMENT

Of all the characteristics of a good leader, the ability to exercise good judgement is the most important. The ability to make the correct decisions, like technical competence, inspires trust and confidence in a group. Good judgement is based on applied intelligence, wide experience, wisdom and the ability to anticipate and analyse. Some leaders seem to be able to make the right judgements while others, even after considerable experience, seem to make perverse decisions. When the going gets tough it is interesting to observe how those who have the ability to make wise judgements and are technically competent assume the role of leader with the tacit consent of all involved.

The factors which go to make up leadership in the second part of this chapter - technical competence, experience, planning and preparation etc. are all concrete, objective skills which can be acquired by individual application. They are not mysterious, abstract qualities reserved for the chosen few. Everybody can improve their leadership qualities, though some may have to work harder than others. These skills are easily discerned and are capable of review like the two earlier models of leadership; through the process of review and evaluation, the skills associated with leadership may be more easily acquired.

1 · 3

Delivering the Expeditions Section

The most effective way of achieving a consistently high quality of adventure, challenge and experience, as well as ensuring the safety and well-being of participants in the Expeditions Section, is through high quality training. This responsibility rests ultimately with the Unit Leader - the person who has the overall responsibility of running an Award Unit. This may be in a school, a Scout or Guide group, a leisure centre, industrial company, youth club, an Open Award Centre or any other body which caters for young people involved in the Award Scheme.

Award Units differ greatly in their structure and form, and there are as many different ways of delivering the Scheme as there are Units within it. Units deliver all four Sections of the Award - Expeditions, Service, Skills and Physical Recreation and, at Gold level, the Residential Project. Adults in the Expeditions Section who are not involved in training, such as some Supervisors and Wild Country Panel members and Assessors, should be aware of the pressures which confront all who work in these Units. It is a great tribute to all those involved in Expedition training in such diverse situations, that they have enabled so many hundreds of thousands of young people to adventure in safety over so many years and that, for the vast majority of participants, the Expedition is regarded as the 'peak experience of their Award.'

A detailed structure of the relationship between Operating Authorities, Award Units and Unit Leaders is set out in the *Award Handbook*. This chapter is concerned with those aspects which directly affect the Expeditions Section, for there are added responsibilities in this Section. All qualifying ventures are of a self-reliant nature. They are not accompanied by Adult Leaders and the participants are therefore totally dependent on their equipment, training and experience for their success and survival. The role of the Instructor is set out in Chapter 6.1. - Training, along with a possible training structure.

Some Award Units consist of one overworked person who has to do everything from running all the Sections, the instruction, driving the bus and acting as the Supervisor during practice journeys and the qualifying venture. Though grossly overworked, these people have one advantage - they are familiar with the total training experience of each individual. Many Units, especially the larger ones, have a team of adults who share different aspects of the training to make the most effective use of their varied talents and provide mutual support for each other. Team work can bring problems as no one person will know how much training individuals have received or whether there have been omissions in the syllabus. Good communication is essential.

Monitoring

One person must have an overall view of the training programme and the participants' progress through the training process. Unit Leaders who are not directly in charge of Expedition training should delegate one person to ensure that the training programme is delivered and that all are adequately trained.

Unit Leaders must ensure that Instructors have the necessary skills, experience and familiarity with the environment in which ventures take place, as well as the skills associated with the mode of travel, to enable them to deliver the Common Training Syllabus (see the *Award Handbook*). More than one Instructor may be needed to deliver the full training syllabus and there are restrictions on who may deliver the first aid training. No specific qualifications are stipulated by the Award Scheme but all Instructors must have the necessary experience and should endeavour to seek appropriate qualifications.

Written records should be kept of the participants' progress against a detailed list of the training syllabus. This is particularly important where the syllabus is delivered by more than one person. It will enable the pages in the *Record Books*, confirming that training has been completed, to be signed with confidence.

A Complete Training Programme

The skills of the Common Training Syllabus and those associated with the mode of travel, usually referred to as the 'hard skills', are clearly defined; the so called 'soft skills' concerned with interpersonal relationships, leadership and team building are less clearly defined, more elusive and frequently overlooked. See Chapter 6.1 - Training. It is very important that the Unit Leader ensures that these aspects of training are fully developed. More ventures fail to reach their destination through a failure in group relationships, a breakdown in morale or a lack of team spirit, than fail through inadequate practical skills.

Groups should be allowed to form naturally from the individual members of the Units and friends should always remain together if they wish. The natural bonding between friends is important since it tends to be more enduring when faced with adversity.

All ventures are a team effort, where the success of the venture is dependent on the success of each individual member and where mutual support and collective experience is essential to achieve a common purpose.

Taking Part

The benefits from a venture are not confined to a successful outcome; the real benefits arise from 'taking part' and from being involved in the process. 'Taking part' is not restricted to the venture itself - it includes the whole of the preparation and training process. It is incumbent on the Unit Leader and the Instructors to make certain that the preparation process for the venture, the training sessions, the practice journeys, the route planning and the team building are all stimulating, challenging, and enjoyable and that every aspect of the training programme has an intrinsic value in its own right for the young people involved.

Realistic Choices

The whole emphasis in this *Expedition Guide* is in making the Expeditions Section more accessible for young people by placing information in their hands so that they may take greater possession of their own ventures, one of the principle aims of the Award.

Joe Cornish

Young people are probably more dependent on Adult Helpers in this Section of the Award than in any other. Vesting choice in young people will involve the Adult Leaders in extra work, but the rewards for the young people are so great that the effort is worthwhile, and there is scope for spreading the workload over other suitably experienced people rather than imposing a greater load on the usual small number of willing helpers.

It is inevitable that a balance must be made between groups' aspirations and the restrictions imposed by the Unit's resources. Many Units will have several groups of participants engaged in the Expeditions Section at Gold level at any one time and there will be other groups involved at Silver and Bronze levels. Few Units, if any, will have the equipment, Instructors, Supervisors and transport to cater for the choices of all the groups undertaking a venture.

Participants have to accept the compromises which will occur when they form groups. Similarly, groups have to accept the limitations which the Unit's resources impose on their aspirations. There still exists considerable opportunity for choice. Groups involved in ventures at the same time may be able to choose the area in which the Expeditions take place, rather than have the choice imposed by the Instructor or tradition i.e. the 'Ford Transit School of Expedition Planning' - groups may carry out their venture anywhere they like but the school minibus is going to the Lakes!

Where the venture area has been pre-determined, Expeditions and Explorations can be undertaken concurrently by different groups, in the same way that Supervisors can look after a number of groups at different levels of Award taking place at the same time. It is not impossible for ventures on foot, on mountain bikes and on water to run concurrently. In Wild Country Panel Areas such as the Yorkshire Dales and the adjacent Pennine Panel Areas it is possible to accommodate ventures at close proximity on foot or mountain bike at Silver and Gold levels and Silver canoeing ventures by canal. This will probably be true of all Panel Areas and some will be able to offer an even greater choice. The careful selection of normal rural or open country will present similar opportunities at Bronze and Silver levels.

There is certainly no excuse for the 'sausage machine syndrome' where groups are issued with the prescribed menu and one of the four prescribed route cards used by all the previous generations of participants and told to 'copy them and give them to the Assessor'! **All groups must devise their own routes and prepare their own route cards as a collective effort involving all the participants within the group.** It is an essential part of the educational process involving navigational skills, personal relationships and team building.

Responsibilities

The responsibilities of the Unit Leader are wide ranging. One of the most important is to ensure that the Instructors who deliver the training programme are suitably experienced and/or qualified and are acceptable to the Operating Authority. It is equally important that they are able to deliver training which is stimulating and enjoyable, and to establish a good working relationship with the young people which will inspire confidence and success.

Safety

The safety of the participants is always of paramount importance. Statistics show that the Expeditions Section has an excellent safety record and it must remain so. There must never be any room for complacency. It is important that the Unit Leader or the delegated Instructor who has overall responsibility reviews, monitors and evaluates the levels of safety at all times. It is most unlikely that injuries will arise from falling down a mountain, or getting caught in a 'stopper' at the bottom of a weir; they are more likely to arise in everyday activities, from walking and riding on the road, or being burned by failing to follow the correct drill when using cooking stoves. It is important that attention is directed to routine practices for they usually contain the greatest hazards.

Responsibility to Parents and Guardians

Parents must be told about the form and nature of the activities in which their children are engaged. This is particularly important for young people who are in *loco parentis*, though the duty of care extends to those above the age of eighteen to the upper age limit of the Award Scheme.

Parents should understand that the nature of journeys within the Expeditions Section is that they are self-reliant and unaccompanied. They should know about the aims and objectives of the Section so that the Unit will have the support of all and the active involvement of many. Parental consent forms should be obtained where necessary. Parents or guardians have a reciprocal responsibility to keep Unit Leaders or Instructors informed of any illness, treatment or physical condition which may be an impediment to participation in a particular activity or activities. Direct communication between the Unit Leader and parents is the best as all who have been involved in education will understand.

Responsibilities to the Participants

The Unit Leader has responsibilities to the participants in ensuring that they are fully and correctly advised. *Record Books* must be cared for and correctly filled in as they represent the validation for all of the activity that takes place under the aegis of the Award Scheme. Signing the pages which confirms the training prior to a qualifying venture is a particular responsibility.

To maintain the integrity of the Award Scheme on behalf of all participants, the requirements and conditions of the Expeditions Section are strictly enforced. All should be aware of the 16 conditions which have to be fulfilled (see Chapter 6.4 - Assessment). Great distress has been caused to both parents and participants, in addition to wasted effort, time, travelling and expense, when ventures have been disallowed because basic conditions have not been observed, for example participants being below the minimum, or above the maximum, age for a Gold qualifying venture.

Literature

It is surprising how many young people complete all levels of the Award without ever having sight of the *Award Handbook*. All participants should have direct access to Award literature so that they can make more informed choices. The books should always be to hand so that they can be referred to.

General Considerations

A number of other considerations require the Unit Leader's attention. While they may be the responsibility of the Operating Authority nothing should be taken for granted, especially as there is such a wide variety of Operating Authorities and User Units in schools, voluntary youth organisations, commerce and industry. It should be ascertained that all legal responsibilities are covered and that the Operating Authorities' regulations have been adhered to. There should be adequate public liability insurance to cover legal liability of themselves, participants and Adult Helpers and damage to third parties as a result of Award Scheme activity. This is a very important consideration for those at the sharp end.

Transport and vehicles are always a concern. Vehicles and trailers must be properly serviced and maintained and comply with EU regulations. Even where restrictions on driving hours do not apply, it would be both foolish and dangerous not to follow current practice in the interests of the participants, parents and the drivers themselves.

Conclusions

The whole of Part 6 - Supporting Roles and Responsibilities and especially Chapter 6.1 - Training, are particularly relevant.

The *Award Handbook* provides general guidance on the role of the Unit Leader and the *Operating Authorities' Guide* contains helpful information.

The *Expedition Guide* expands the information given in the Expeditions Section of the *Award Handbook*, interprets the requirements and conditions in the light of collective experience and provides general advice, but the final reference source for the requirements and conditions must always be the *Award Handbook*.

The Process

Initial Briefing and Training	Participants should plan a venture in terms of purpose, mode of transport, chosen environment etc. and undertake appropriate training.
Practice Journeys	Participants must undertake sufficient practice journeys to enable them to complete their planned venture.
Qualifying Venture and Debrief	A qualifying venture with a purpose must be undertaken, with a group of between four and seven young people, followed by a debrief.
Review	All participants should present an account of their journey related to the purpose.

EXPEDITION GUIDE

THE DUKE OF EDINBURGH'S AWARD

Part
2

David Noton / National Trust

OPTIONS

AND

ALTERNATIVES

Part 2

Options and Alternatives

Introduction and Requirements

There are three alternative forms of venture within the Expeditions Section:

- **Expeditions** which have journeying as their principal component.

- **Explorations** which involve less journeying and a greater proportion of time spent on approved first-hand investigations or other specified activities. Explorations must take the form of a journey and involve a minimum of ten hours' travelling time.

- **Other Adventurous Projects** which are of an equally, or more demanding nature but depart from the specified requirements or conditions. This option is available at Gold level only.

A number of options are available, such as ventures abroad, for participants to extend their opportunities and facilitate the completion of the Expeditions Section. The *Award Handbook* differentiates between **'requirements'** and **'conditions'.**

Requirements apply to all three alternative forms of venture at Bronze, Silver and Gold levels and to all journeys regardless of the mode of travel.

Conditions apply specifically to the level of the Award - Bronze, Silver or Gold - the mode of travel and the environment in which the venture must take place. These are set out in the relevant chapters of this *Expedition Guide*, particularly in Part 4 - Advice and Skills Associated with the Mode of Travel.

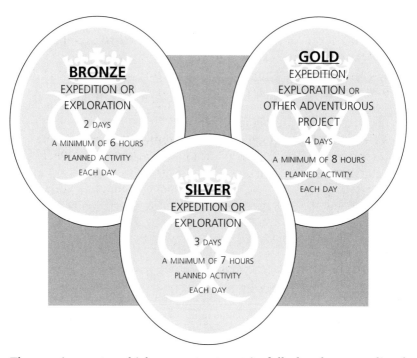

The requirements, which are not set out in full elsewhere, are listed below.

All ventures must comply with the following:

- All ventures must take place in the context of an unaccompanied, self-reliant journey in the countryside, or on water, conceived with a purpose and undertaken by the participants' own physical effort, without motorised assistance and with the minimum of adult intervention.
- All qualifying ventures must have a clearly defined purpose.
- After completing the venture participants must produce a report related to this purpose.
- Ventures should take place between the end of March and the end of October.
- Ventures should involve joint planning and preparation by all the members of the group.
- Groups must consist of between four and seven young people.

- Accommodation will be by camping.
- Clothing, footwear and equipment should be suitable for the activity and the environment in which it is to be used and generally conform to current accepted practice. The equipment must be capable of resisting the worst weather since, in the event of a serious deterioration in conditions, safety may well depend on its being able to withstand the prevailing conditions.
- All equipment and food to be used during the venture must be carried by the group. Participants must always carry the emergency equipment listed in Chapter 3.1 and their own sleeping bag. Groups must always carry at least one tent, cooking equipment and some food so that they always have the shelter of a camp to fall back on if they should fail to reach their destination.
- Participants must be trained in the necessary skills be experienced Instructors.
- Participants must undertake practice journeys to ensure that they are able to journey safely and independently in their chosen environment.
- All ventures must be supervised and assessed by suitably experienced adults.
- Groups must not include a participant who has successfully completed a qualifying venture at the relevant level or at a higher level of the Award.

These requirements, coupled with the conditions listed under the various modes of travel, are the foundations for building a safe and successful venture which will comply with all the demands of the Expeditions Section of the Award.

In exceptional circumstances it may be necessary to vary the Award's requirements to meet the specific needs of certain individuals or groups. For example, it may be necessary to use barns, bothies or mountain huts instead of camping, or individuals may not be able to carry a full set of equipment, and require non-essential equipment to be pre-positioned at the camp sites. Approval for such variations must be given by the appropriate person within the Operating Authority as Operating Authorities are responsible for the quality of the participants' experience and their safety.

Joe Cornish

Expeditions

The word 'Expedition' is used in two ways within the Award. The words 'Expeditions Section' are used to differentiate the journeying Section from the Service, Skills and Physical Recreation Sections and the Residential Project. It is frequently used in an all-embracing form to include Explorations and Other Adventurous Projects as well as the traditional journey measured in distance or travelling time, because all ventures involve a journey.

In this chapter, 'Expedition' is used with its specific meaning to describe the traditional journey with a prescribed mileage or a mandatory number of hours of travel on each day of the venture. This differs from Explorations and Other Adventurous Projects which are discussed in Chapters 2.3 and 2.4.

JOURNEYING

Ventures must be by the participants' own physical effort and, while this imposes some restriction on the method of journeying, there are more than sufficient modes of travel available to meet the aspirations of all participants and provide opportunities for the widest range of purposes which form the foundation of a venture.

Expeditions on land must take place in normal rural or open countryside, or in one of the designated Wild Country Areas avoiding villages and other habitation where possible and may be:

- On foot.
- By cycle.
- By horse.

Expeditions on water must be by:

- Open canoe or kayak.
- Rowing/pulling boats.
- Sailing dinghies, keel boats or yachts up to a prescribed length and number of berths.

The Award Scheme has gone to considerable lengths to encourage participants to be more adventurous in choosing alternative forms of venture and different modes of travel. In spite of all these efforts the predominant form of venture continues to be an Expedition on foot. This is largely due to the lower costs of foot ventures, where the cost of expensive equipment such as canoes, sailing dinghies, bikes or horses can be avoided. There is no doubt that many young people and adults involved in the Award Scheme enjoy the objectivity of Expeditions with their minimum mandatory distance or prescribed number of hours of travelling.

Journeying is the principal component of Expeditions and the minimum distances, or the minimum number of hours of travelling, for the various modes of travel are set out below:

Requirements for qualifying ventures

LEVEL	BRONZE 2 days, 1 night	SILVER 3 days, 2 nights	GOLD 4 days, 3 nights
Hours of planned activity	Minimum of 6 hours each day	Minimum of 7 hours each day	Minimum of 8 hours each day
Foot Expeditions	24km/ 15 miles	48km/ 30 miles	80km/ 50 miles
Cycling/Horse Riding/ Canoeing/ Rowing/ Expeditions	At least 4 hours journeying each day	At least 5 hours journeying each day	At least 6 hours journeying each day
Sailing Expeditions	12 hours planned activity over the 2 days	21 hours planned activity over the 3 days	32 hours planned activity over the 4 days
Explorations (all modes of travel)	At least 5 hours journeying over the two days	At least 10 hours journeying over the three days	At least 10 hours journeying over the four days

PURPOSE

All Expeditions must have a clearly defined purpose and a report, related to the purpose of the venture, must be presented on completion of the journey.

The difference between the Expedition and Exploration options is only one of emphasis between the journeying and the discovery element.

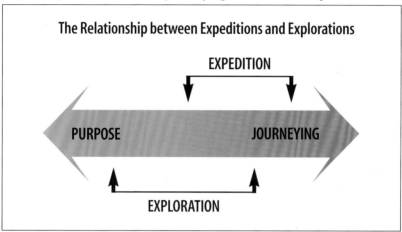

The Relationship between Expeditions and Explorations

EXPEDITION

PURPOSE   JOURNEYING

EXPLORATION

From traditional Expeditions, where journeying is the principal component, vast numbers of reports of outstanding merit have been produced which have demanded depths of scientific and environmental awareness, artistic perception, scholarship and an understanding of personal and group relationships.

The foundation of a successful Expedition is a carefully chosen purpose for the venture which, coupled with appropriate forms of observation and recording, will result in a report reflecting the value and significance of the venture.

EXPEDITIONS USING MORE THAN ONE MODE OF TRAVEL

The safety record of the Expeditions Section is based on the essential simplicity of the ventures with the training directed to the mode of travel, the environment and instruction which enables participants to be self-reliant in unaccompanied situations. This, and the fact that all group members have their clothing, food, camping and emergency equipment with them to fall back on if things go wrong, ensures a high level of safety. These standards must be maintained.

A very small number of Expeditions submitted for validation have involved different modes of travel during the same venture. For example, one venture involved a day's walking, a day's cycling and a day's canoeing. The Award views these Expeditions with concern because they:

- Erode the safety margins.
- Impose an extra training workload on the participants and their Instructors and involve extra cost.
- Increase the group's dependence on adult participation and the amount of intervention by Adult Helpers.
- Result in a contrived nature to the ventures.
- Have a diminished regard for the spirit of the Expeditions Section of the Award.

In the example quoted above participants would have not only to complete the full wild country training syllabus but also to complete the swimming requirements and the appropriate BCU level of proficiency, in addition to attaining the Cycleway Certificate. This would nearly treble the demands of the syllabus and necessitate three times as many practice journeys to meet the conditions set out in the *Handbook*. The training standards are minimum standards designed to protect the participants, the Operating Authorities and the Award Scheme which is not prepared to accept any diminution of these standards.

All food, clothing, camping and emergency equipment must be carried for the whole of the journey. It is possible to borrow or hire cycles and canoes; it is much more difficult to obtain the panniers and waterproof containers without which cycling and canoeing ventures cannot take place.

Where routes are dictated by the need to change from one mode of travel to another rather than determined by the purpose of the venture, there is nearly always an increase in the use of roads and centres of population rather than being remote from habitation.

Such ventures also place an unfair burden on Assessors and in particular on the Wild Country Panels which need to provide Accredited Assessors with the necessary qualifications or equivalent experience in the various disciplines involved in the different modes of travel.

Tony French (Halina/Fuji Bursary)

The Award Scheme always wishes to encourage initiative and to respond to the enthusiasm of the participants. Some ventures involving more than one mode of travel have been successfully planned in the past. Participants have carried canoes up to Lakeland tarns to carry out depth soundings as part of an Exploration. Portage, which involves carrying canoes between navigable water, is a recognised Expedition technique providing, of course, that participants do it themselves without motorised assistance.

While the conditions do not expressly forbid journeying by different modes of travel during the same venture, considerable thought should be given at the planning stage to any such project to ensure that the venture meets not only the requirements and conditions, but the spirit of the Award as well.

Joe Cornish

Explorations

Explorations offer an exciting alternative to the more usual Expedition with its fixed distance or hours of travel, and enable participants to place a greater emphasis on the 'discovery' element rather than on the 'journeying'. The difference between an Expedition and an Exploration is only one of balance between these two aspects.

"An expedition is a journey with a purpose
An exploration is a purpose with a journey." (A.P. Lowery)

The traditional Expedition with its mandatory distance or hours of travel, whether in wild country or on water, makes heavy physical demands upon the participants, and though Expeditions must always have a purpose and a discovery element, the amount of time available is frequently restricted by the demands of the journey. Explorations must always take the form of a journey, but the journey time is reduced to a minimum of five hours of journeying over two days at Bronze level and ten hours of journeying over three days and four days respectively at Silver and Gold.

Though the journeying time is reduced in Explorations, it is still important to provide the separation from habitation which is so essential in bringing about a true sense of adventure and sustained endeavour, and for the outcome of the venture to be in the hands of the participants alone.

A MENTOR

Adults involved in Explorations can best help the participants by enabling them to clarify the aims of the Exploration and focus on realistic and achievable tasks, suggesting sources of information and suitable areas for the venture, as well as giving advice on the presentation of the final report. It is helpful, as the participants' ideas take shape, that someone who is knowledgeable and experienced in the chosen field of interest, and who is willing to help, is involved as the group's Mentor. The Mentor is someone who combines the function of advisor, guide and friend and helps the group to derive the greatest value from the purpose of the venture. A Mentor should be involved at the earliest stages of the planning so that the participants may avoid following unproductive avenues of study and have a viable venture with a purpose which may be achieved.

CONDITIONS

Distance

The Exploration must take the form of a journey. As with all ventures, the minimum of five or ten hours' travelling commences when the group reaches the venture area. It is intended that the minimum of five or ten hours' travelling time should enable the group to achieve the necessary remoteness in terms of time and distance to make them self-reliant and dependent on their own resources. The five or ten hours of journeying is a minimum base line; the Exploration option provides a great deal of flexibility in the journeying and the purpose of the Exploration.

Environment

Bronze

The environment must be appropriate to the venture and **unfamiliar** to the participants. The venture should take place in normal rural country avoiding the use of roads where possible, and making every effort to avoid villages.

Silver

The environment must be appropriate to the venture and **unfamiliar** to the participants. At Silver level normal rural or open country may be used, with the use of any minor roads or byways limited to that which is necessary to move between areas of open country. The conditions exclude all use of cities, towns and villages as sites for Exploration or accommodation, with the venture taking place sufficiently removed

from habitation to ensure that the group is self-sufficient and dependent on its own resources. **For Explorations at Silver level taking place in wild country, the equipment, training and preparation must always be to the standard required for Gold ventures.**

Gold

Gold Explorations must take place in wild or open country remote from habitation. The environment must suit the purpose of the venture and be **unfamiliar** to the participants, with the same restriction on the use of roads and areas of habitation as at Silver level.

The mandatory requirements for the use of wild country for ventures on foot, cycle or horse apply at Gold level. To increase the scope for Explorations open country, estuaries, marshes, fens and coastal areas which are **remote from habitation and tourists** may be used, provided the conditions set out above are followed.

Explorations on Water

Explorations afloat must be on water appropriate to either the Silver or Gold level as set out in the *Handbook* for the craft being used, with the same levels of qualification or competence being required. The conditions exclude all use of water in or near to cities, towns and villages as sites for Exploration or accommodation.

Accommodation

Accommodation will be by camping. Normally a different camp site should be used each night so that the Exploration takes the form of a journey. Where the nature of the Exploration necessitates the same camp site being used on more than one night, this is permissible, but travelling around the camp site may not be included in the ten hours of travelling time.

Mode of Travel

Any form of travel which may be used in the Expeditions Section may also be used for Explorations. While the majority of Explorations take place on foot, the use of cycles or horses extends the area in which Explorations can take place and the use of permitted forest tracks and trails are ideal for these modes of travel, as are bridleways which may be legally used by cyclists.

Glyn Satterley. Still Moving

Ventures afloat open up new and exciting possibilities for Explorations. Canoes, sailing dinghies and rowing/pulling boats are ideal vehicles for Explorations, providing excellent mobility as well as the capacity to carry Exploration equipment. The open canoe has advantages over the kayak in carrying capacity, but kayaks have been carried up to Lakeland tarns for Explorations involving depth soundings.

If the mode of travel is being changed from that of previous ventures, extra time is nearly always needed to gain the necessary competence in the mode of travel.

OTHER CONDITIONS

The requirements and conditions set out in the *Award Handbook* all apply to Explorations. Participants must be equipped to the standards set out in the *Handbook* and the *Expedition Guide,* and the full syllabus of training for the level of Award must be undertaken, along with at least the minimum number of practice journeys.

Because all Explorations take place away from habitation, it is essential that the participants are trained to cope with the particular environment and are aware of any hazards which may be associated

with it. Marshes, estuaries and the seashore have hidden dangers which may not be apparent to those who are not very familiar with these environments. Only those accustomed to coastal conditions know how easily it is to be trapped by the tide, and experienced cavers appreciate how quickly flood waters can rise. **Crags, caves and old mine-workings pose special problems requiring considerable expertise on the part of both the participants and the Supervisor. These places are best confined to those who have competence in these activities.** Groups will be working alone expect for the daily visit of the Supervisor/Mentor and the Assessor.

Supervision

During the acclimatisation period and the qualifying venture, the group will be supervised on behalf of the Operating Authority in the same way as with all other ventures. A Mentor, suitably experienced in the skills of expeditioning, may act as Supervisor with the approval of the Operating Authority.

Assessment

Explorations are assessed in the field in the normal way by an Assessor who will ensure that the conditions of the Section are fulfilled and then complete and sign the appropriate page in the *Record Books*. The participants may submit the report to any adult who has been actively involved in the venture or its preparation, such as the Assessor, Supervisor or Instructor. With Explorations, an obvious person to whom the report could be submitted would be the Mentor, if one was involved. The Assessor should check on who will receive the report. On receipt of a satisfactory report, the other relevant page of the *Record Books* should be signed by the person receiving the report. The use of a Mentor, with expertise in the particular study or purpose of the Exploration, simplifies the assessment considerably and removes many of the problems which have occurred in the past. It reduces the work-load of the Assessor in over-used Wild Country Panel Areas and ensures that the participants' efforts receive the attention and appreciation which they usually so richly deserve.

ENSURING SUCCESS IN EXPLORATIONS

To ensure a successful outcome to an Exploration, the conditions set out above should be studied carefully and an Exploration designed to comply with these requirements.

There is an endless choice of subjects for Exploration. The Duke of Edinburgh's Award published the *Exploration Resource Pack* in 1994. This publication not only provides inspiration, but suggests ideas and ways of approaching the venture. It includes:

- Setting up the Exploration.
- Ensuring that it conforms to Award requirements.
- Examples of good practice.
- Ideas for projects - how, why, what and when.
- Planning hints - general guidelines, source material, reference material.
- Simple home-made equipment.
- Practical methods.
- Observing and recording techniques - photography etc.
- Presentation, results, conclusions and style of the report.

There is an endless variety of subjects which can form the basis of an Exploration, and they do not have to be based on the sciences or field study type of projects. Arts and the humanities are equally important and the scope is only limited by the participants' initiative and imagination.

There is no need for groups to be 'academic' to take part in an Exploration. There is ample opportunity for all to participate at every level of academic ability or development. Most field studies are built around simple observations, counts and measurements, using simple home-made equipment and there are no restrictions on the way in which the Exploration may be presented.

Explorations should be based on practical first-hand studies, artistic or aesthetic expression or recording, carried out in the field during the course of the venture. The studies, if not of an original nature, should be concerned with the gathering of material which is new to the participant. The journey and the subsequent investigations or recordings are essential to the project. If the information could be obtained from tourist guides or the local library, there would be no point in carrying out the journey in the first place and so the venture would be irrelevant. It is important, however, that all field work should be based on a study of relevant material and literature carried out beforehand.

Because the conditions exclude all cities, towns, villages and habitations for the Exploration or accommodation during the venture, any research which involves visiting a town or village must take place beforehand or during the acclimatisation period. For example, a literary study of Wordsworth or the Bröntes would probably involve a visit to Dove Cottage or the Haworth Parsonage. The visit to these overcrowded locations could take place during the familiarisation period of forty-eight hours immediately prior to the venture; both are located within designated Wild Country Areas.

During the planning and preparation of the Exploration, participants, Supervisors and Mentors should regularly check progress against the criteria listed below to ensure the validity of the venture. The Assessor will be using the same criteria. Participants should decide early in the planning stage who will receive the report and enter this on the *Expedition Notification Form* before it is dispatched.

CRITERIA

Sixteen criteria are listed here to assist in the preparation of an Exploration:

1. The Exploration must be located in an area where it is safe for the participants to work by themselves, taking into account their training, preparation and experience.
2. The venture area should be sufficiently remote from habitation for the group to be self-reliant and dependent upon its own resources.
3. The venture must take the form of a journey with an appropriate balance between project work and journeying (minimum 10 hours) during the qualifying period.
4. The Exploration must be suited to the aptitudes and abilities of the participants.
5. The purpose of the Exploration should be limited and properly focused.
6. It should be possible to fulfil the purpose of the Exploration in the time available.
7. Fulfilment of the purpose should depend upon the journeying element; if it can be completed by visiting a library or museum, it is not suitable for an Award Exploration.

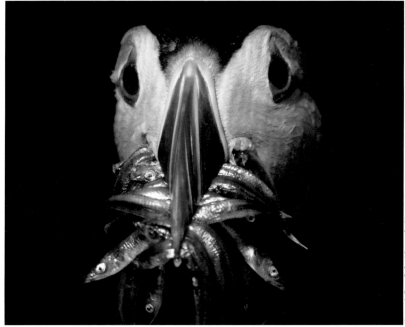

Kenneth Day, Oxford Scientific Films

8. The Exploration must be based on the participants' first-hand observations, experience or study.
9. The recording techniques and equipment should be appropriate to the study.
10. The fieldwork during the Exploration should be based on previous study or activity.
11. It is useful if someone with specialist knowledge, a Mentor, is involved from the early planning stage of the venture.
12. The format of the report should be agreed beforehand.
13. The group should be aware of the conservation issues and should not harm that which it is studying.
14. If the Exploration involves a more hazardous pursuit such as a water-based activity, climbing or caving, the participants must have the necessary skills and, above all, the experience to cope with situations which may arise.
15. The Supervisor should ensure that the participants are aware of any potential hazards which the Exploration may present.
16. The role of the Mentor, Supervisor and Assessor should be agreed beforehand and adequate lines of communication should be established prior to the venture.

If the proposed Exploration is tested against these criteria in the early stages of planning and during the subsequent development, it should ensure a successful and rewarding outcome to the venture.

Other Adventurous Projects

Other Adventurous Projects are of an equal or more demanding nature than the normal Gold Expedition or Exploration but depart from the specified conditions of the Expeditions Section. Because of their special nature Other Adventurous Projects are restricted to Gold level. They provide participants with even greater scope for imagination and enterprise for their venture and many distinguished enterprises have taken place. They include a wide range of ventures such as crossing the Atlantic in a yacht, the ascent of Himalayan peaks, journeys through desert and jungle and exciting, unusual ventures in the United Kingdom - the list is endless. Other Adventurous Projects must still comply with the spirit and philosophy of the Expeditions Section - an unaccompanied journey where the participants are self-reliant and sufficiently remote from habitation and dependent on their own resources, travelling by their own physical effort and without motorised assistance. Sadly, in recent years, a very modest stream of distinguished Adventurous Projects has become a flood of mediocre submissions, many of which could not be justified as normal Expeditions or Explorations, let alone as Other Adventurous Projects.

The Panel which considers Other Adventurous Projects (OAP) receives submissions which are frequently an abuse of this type of venture. Some of the more common abuses are listed below:

- Many groups choose to do an Other Adventurous Project to avoid specific conditions of the Expeditions Section which they find inconvenient.
- Many foreign projects, while very worthy in themselves, are not appropriate for the Expeditions Section and would be more suitable if they were submitted as Service Section projects or for the Residential requirement.
- Projects are initiated by leaders or staff to satisfy their own needs. They wish to go caving in Korea, so they use the Award Scheme as a justification, a means of obtaining sponsorship and of drumming up financial support and equipment.
- A number of projects are enterprises marketed by commercial organisations which are primarily concerned with filling seats on safaris, or increasing the numbers engaged in their ventures.
- Participants become involved in other people's projects rather than initiating their own. They remain 'also-rans', never being involved in the initiation, planning, preparation or major decision-making.
- Young people take part in ventures for which they do not have the relevant skills, or their skills are not adequate. They are either exposed to unjustifiable risks or they are subjected to such close adult supervision that it destroys the purpose and spirit of the venture.
- Some engage, one suspects, in an Other Adventurous Projects 'just to be different' believing, quite wrongly, that it has greater status than the more usual Expedition or Exploration.

To eliminate these negative aspects and improve the quality of submissions, the Award Scheme issues the following advice on Other Adventurous Projects.

Choice: The essential simplicity of the traditional Expedition and its role in personal and group development is such that any departure from this format should not be undertaken without careful consideration.

Age: Other Adventurous Projects are best suited to young adult participants in the Award - those between the ages of 18 and 25 and who are not in *loco parentis.*

Initiative: Other Adventurous Projects should be initiated, planned and prepared by the participants themselves and be their 'brainchild'. The Projects should be distinguished by boldness, imagination and enterprise. A project which only modestly exceeds the expectations of a Gold venture, but which is entirely conceived and executed by the participants, is preferable to a much more ambitious scheme which has been initiated and planned by Adult Leaders.

Conditions: Other Adventurous Projects must take place within the context of an Expedition. Reading the requirements and conditions in the Expeditions Section of the *Award Handbook* should always be the first step. The conditions which define the environment for Expeditions - wild country, isolation, remoteness from habitation and the need for dependency on one's own resources - are particularly important ingredients for a successful project.

Appropriate training for the mode of transport, thorough preparation and completion of the appropriate number of practice journeys are essential for Other Adventurous Projects whether in the UK or abroad.

Involvement in the customs, culture and life of overseas countries is vital in any visit, and time should be set aside for this involvement, but always before or after the Adventurous Project itself. The Award frequently gives Other Adventurous Project approval for a specific part of a much longer project.

Skills: Participants themselves must have the necessary level of skills and the techniques to be able to cope with any demands made upon them; their safety must not be dependent on adult intervention.

Joint Initiative: An increasing number of Youth Organisations and Operating Authorities run events to all parts of the world, involving many young people in social and cultural exchanges. It is acceptable for Award participants to plan and carry out Expeditions and Explorations within the context of these major events. Providing they comply with the requirements, are properly supervised and assessed and are approved by the Operating Authority, it does not matter whether the venture is on foot in the Pyrenees or by dug-out canoe in the Amazon Basin. If, however, participants wish to carry out an Other Adventurous Project during one of these events, the two conditions listed below must be observed and prior approval obtained from Award Headquarters.

Joe Cornish

Submissions should be made by the participants themselves after reading these notes carefully together with the advice given in the Expeditions Section of the *Award Handbook*.

All submissions should be made via Operating Authorities in sufficient time to reach the National Award Office AT LEAST TWELVE WEEKS prior to the date of departure.

The Award cannot accept any responsibility concerning the suitability of the venture for the participants concerned, all safety aspects, the adequacy of the training or emergency procedures. These responsibilities must rest with the Operating Authority who should state that they approve of the venture.

Solo ventures would not normally constitute an Other Adventurous Project, but any truly enterprising venture will be considered providing that the Operating Authority has approved and accepted the responsibility for the venture.

In giving approval for for Other Adventurous Projects, Award Head Office only confirms that the project is acceptable as a Gold qualifying venture within the Expeditions Section if it is successfully completed.

Proposals must state how all aspects of the venture will meet the requirements of the Award. Full explanations under the following headings will ensure a speedy consideration of an application:

1. The purpose of the venture.
2. The reasons why the venture will differ from the requirements and conditions of a standard Gold qualifying venture and why it should be considered as an Other Adventurous Project.
3. The location and nature of the environment in which the qualifying part of the venture will take place.
4. The mode of travel.
5. The nature of the accommodation and the catering arrangements.
6. The duration of the qualifying part of the venture, with dates if possible.
7. The number involved in the venture with the names and ages of those wishing to use it for their Award.
8. The experience and skills of those involved which would justify their undertaking the Other Adventurous Project.

9. The planning, preparation and training, including any specific aspects related to the nature of the venture. Details of practice journeys.
10. Any special precautions or procedures in case of an accident or emergency.
11. How the proposed venture is to be supervised and assessed, and who will receive the presentations.
12. The proposed departure date from home to the location where the venture will take place.

If this advice is followed it should significantly reduce the amount of letter writing, the number of telephone calls and smooth the path of the proposed venture, enabling memorable Other Adventurous Projects to take place anywhere in the world.

Ventures Abroad

An increasing number of ventures are taking place abroad, especially in continental Europe. Groups from the South of England find it no more difficult or expensive to travel to the Continent than to the North of the British Isles. All ventures abroad have to comply with exactly the same requirements and conditions as those which take place within the United Kingdom, i.e. training, practice journeys, the type of country (normal open or wild country), distance and duration. The supervision and assessment of these ventures must all conform to normal practice.

Ventures on the Continent are not as easy to arrange or execute as those in the United Kingdom, even after allowing for the integration of Britain into the European Community. Ventures elsewhere in the world continue to present a considerable challenge to the participants. The difficulties facing those who venture abroad seem to serve as an extra challenge which increases the determination of both the participants and the adults involved.

This chapter is directed towards continental Europe where the vast majority (about 85%) of ventures abroad take place. It will serve as a basis for planning all ventures abroad, though ventures outside the European Community will always involve extra preparation, precautions and procedures, ranging from passports, visas and vaccinations to import and export documentation for Expedition equipment.

Ventures in continental Europe open up enormous opportunities for challenge. The lower densities of population in the rural areas and the sheer physical size of the terrain make it possible to achieve the isolation and remoteness from habitation which is so intrinsic to Award ventures, without having to resort to high altitude in mountainous areas. Access to the countryside is generally much easier than in the UK, though large tracts of country are restricted by hunting and shooting, which is much more widespread on the Continent, and camp sites may be harder to come by in certain countries. Mountain bikes are widely used in many areas with far fewer restraints on their legitimate use than in Great Britain.

The rivers and canals on the Continent offer splendid facilities for canoeing and rowing ventures. Their impressive length, breadth and volume open up a new dimension for British canoeists and rowers, with endless miles of BCU Grade 1 and 2 water from which to choose. The sailing waters of northern Europe and Scandinavia have hardly been broached by dinghy sailors within the Award Scheme.

All visits abroad provide wonderful opportunities for a stimulating diversity of activity in addition to the qualifying venture and the vital acclimatisation period. Time should obviously be set aside for cultural and social exchanges and the experiences which characterise and distinguish nations and cultures. There should always be opportunities to visit and view features and aspects of the country or area which give it its distinctive national character, but these activities should always take place **before** or **after** the venture because they are precluded by the conditions from taking place during the venture itself. The ventures themselves provide excellent opportunities to study the geographical, geological and other features which distinguish and characterise places in different parts of the world. The purpose for the venture should not, however, be based on comparing cultures and lifestyles as these studies involve people and inhabited places which are excluded during ventures in the Expeditions Section.

The following advice is offered on ventures abroad, not to discourage but to ensure that those who venture abroad enjoy the same levels of safety and well-being as those who carry out their Expeditions in Great Britain.

It is essential that all groups and their Leaders familiarise themselves with the customs and practices of the countries which they intend to visit, so that the standards of behaviour always conform to the expectations of the host country and do not give offence, and that the reputation of the Award Scheme is maintained.

All participants should be sufficiently familiar with the host country's language, not only to exchange the time of day and common courtesies, but to use the telephone effectively in an emergency. This will involve practising on the telephone during the familiarisation period, so that all are familiar with dialling tones, procedures and coinage.

Laws and regulations concerning travel, vehicles, supervision and even the activities vary between one country and another. Leaders need to familiarise themselves with the relevant regulations and practices.

The vast majority of wild country areas on the Continent have no mountain rescue or search teams. Groups, Supervisors and Assessors will be entirely dependent on their own resources. Some areas, such as the Alps, have highly efficient professional rescue teams but their services are very expensive, costing many thousands of pounds which will always have to be paid for, usually before the group is able to leave the country. Insurance must cover search and rescue costs as well as the repatriation of the sick or injured and the need of **both** parents to visit hospitalised participants abroad.

In the case of serious accidents, the restraint which the police exert in the UK over the release of names to the press until next of kin have been informed does not apply abroad. There is a strong probability that parents may find out about a serious accident by TV, radio or press unless an efficient two-way communication between the group and parents has been established beforehand for use in an emergency.

Temperatures well above those usually experienced in the British Isles during the summer months increase the probability of exercise-induced heat exhaustion and a greater need for protection against the sun and dehydration. Thunderstorms and lightning strikes are more severe and frequent, and the wildlife may pose a greater threat through rabies and snakebites.

Gold Expeditions and Explorations on foot, cycle or horseback must take place in an agreed Wild Country Area. Examples in Europe are the Alps, the Pyrenees, the Massif Central, the Cévennes, the Ardennes, the

Harz Mountains and southern Bavaria. Only some of the Wild Country Areas have been designated and other areas will be included as demand necessitates. The examples listed above will give an indication of the types of country involved.

In the British Isles the use of long distance footpaths is not acceptable for Expeditions other than to link two areas of terrain together. On the Continent, a more permissive stance is taken and may be allowed in certain situations where long distance paths are the only way of obtaining access to large tracts of land, or making progress through the very difficult terrain or forests. By careful selection of the area it is possible to use long distance paths which are remote and rarely used, where the very vastness and remoteness of the landscape can be awe-inspiring to those who have only ventured in England or Wales. Some of these paths present a serious challenge in themselves.

The need for a prior visit or reconnaissance of the intended venture area by participants, Leader, Supervisor or Instructor cannot be overstressed, though some groups have been successful by taking the advice of a person who is based in the area. Large areas of the Alps and Pyrenees present such difficulties and hazards that they are beyond the reach of all but a few groups of the older and stronger participants. The foothills of these regions provide more suitable areas for backpacking, and the advice to **travel through wild country rather than over** is never more appropriate. Many groups will have great difficulty in completing the 80 kilometres in the mandatory 4 days, even in the foothills of these areas, unless they are chosen with great care.

Ventures which necessitate around 500 metres of ascent per day, or sometimes much less, may present exhaustion problems, especially in southern Europe during the summer months. If it is unrealistic for 80 km of backpacking to be completed in the prescribed four days, then the terrain is far too difficult and should not have been chosen in the first instance. It is a hard lesson for the participants to learn and represents a sad waste of effort, time and money.

Ventures on the Continent require long preparation and those in other parts of the world even longer. Planning, training and preparation usually need to start, at the latest, in the Autumn prior to a Summer venture. This commences with the participants, Instructors and Supervisor preparing an all-embracing schedule and checklist, which will include travel arrangements and bookings, and end with a pre-departure meeting two or three days before leaving, when all checks will be completed. All documentation should be collected and safely stored to avoid last minute panics. The checking process should always be undertaken by two people.

Most groups travel by road and it is essential that vehicles comply with the appropriate regulations, are serviced prior to departure, insured for breakdown and the repatriation of vehicle and passengers. Drivers must not only observe Community directives on driving times but must follow a rest routine which will ensure the safety of all, in what is probably the most dangerous aspect of the venture.

As with all ventures in the Expeditions Section, parental consent should always be sought for those under 18. Parents/guardians should be kept fully informed of the nature of the venture, how it will be supervised, and the arrangements made for the participants during 'free time'.

Operating Authorities usually have their own Codes of Practice and regulations which must be observed.

MAPS

Most of western Europe is mapped at the 1:50 000 scale and many countries such as France, Germany and Switzerland are also mapped at the 1:25 000 scale. Foreign maps may be purchased in this country and a wide range of western European maps of the more popular areas are available off the shelf. Many foreign maps do not have the familiar grid lines and, in many cases, not even the lines of latitude and longitude. Participants will have to draw their own, based on the reference numbers on the edges of the maps using a sharp soft pencil. This should be done while the venture is being planned so that participants, Supervisors and Assessors are all using the same reference system and have similarly amended maps.

Route tracings should be from either the 1:50 000 or 1:25 000 maps after consultation with the Assessor and the Supervisor. They should include full details of the map(s) and scale being used for both the tracings and the venture. If not obvious, tracings must include information on how the grid references were determined. If necessary measure the distance in millimetres from the left edge of the map towards the right to give Eastings, then measure the distance North from the bottom edge to give Northings. Sometimes the best solution will be to provide the Assessor with a prepared map.

Compass variation in western Europe will be minimal during the life of this *Expedition Guide* and may generally be ignored, though it may be prudent to check for the area in which the venture takes place. Projections other than the Transverse Mercator used in our Ordnance Survey may be involved, so ensure that the left- and right-hand edges of the map run North and South. For additional information refer to *Land Navigation, Route Finding with Map & Compass* published by The Duke of Edinburgh's Award and available through Ordnance Survey stockists.

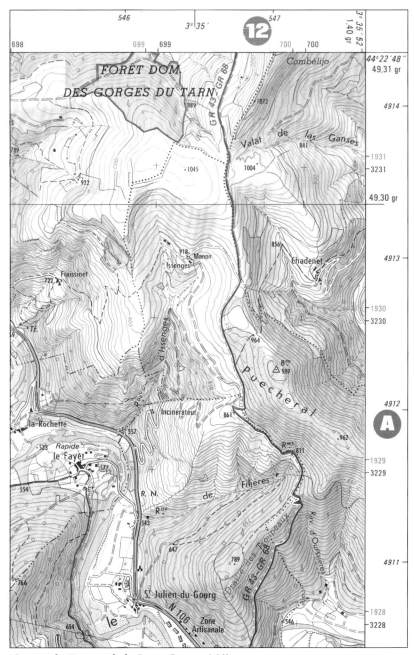

Gorges du Tarn et de la Jonte Causse Méjean

- Carte no.2640 OT
- © IGN Paris + 1992 édition
- Autorisation No. 90-6082

THE NOTIFICATION PROCESS

A procedure has been established for the notification of all Award ventures abroad.

The regulations and conditions for all Award ventures abroad are exactly the same as for those in the United Kingdom.

Those wishing to carry out their ventures outside the United Kingdom must have their ventures approved by their Operating Authority which has the legal responsibility for their safety. Three copies of the *Notification Form for Ventures Abroad*, the 'blue form', which is available from Award Secretaries and Regional Officers, should be completed, along with tracings and relevant details. Two copies should be sent to the Operating Authority for approval **at least 12 weeks prior to departure.** Once the venture has been approved by the Operating Authority a signed copy should be sent to the Award Secretary/Regional Officer for information. The Award Secretary/Regional Officer will allocate the venture a number and inform the group or unit. This number should be entered in the participants' *Record Books* following the successful completion of the venture.

Unit/Group

- Complete 3 forms, sets of tracings and venture details.
- Keep 1 form, tracings and details.
- Send 2 forms, tracings, maps and details to the Operating Authority at least 12 weeks in advance for approval.

Operating Authority

- Approve the venture and confirm that the Supervisor is suitably experienced and/or qualified to ensure the safety of the participants.
- Ensure the venture conforms to the criteria listed in the *Award Handbook.*
- Appoint an Accredited Assessor or approve a suitable independent Assessor.
- Sign BOTH forms.
- Keep 1 form, tracings, maps and details.
- Submit 1 signed form with tracings and details to the appropriate Award Secretary or Regional Officer for notification and to receive anotification number.

Award Secretaries / Regional Officers

• Allocate and send notification number to Unit or group. Keep notification form, tracings and details. Submit statistics to Award Headquarters annually on request.

The notification number is constructed as follows:

1. The initial letters of the Territorial or Regional Office e.g. NI, WA, SC. (Northern Ireland, Wales, Scotland); SW, NE etc.
2. The sequential number of the application for a venture abroad in the current year.
3. The year.
4. The first three letters of the country in which the venture is to take place e.g. FRAnce, GERmany, AUStria. The probability of confusion is not very great even though AUS might refer to Australia.
5. The last factor is the region or area in which the venture takes place e.g. ARDennes, VOSges, CEVennes, PYRenees, HARz.

> **VENTURE NUMBER   SE/3/97/FRA/ARD**
> **Region/3rd Notification/Year/Country/Region**

The notification number has three purposes: to authenticate the venture in a similar manner to Gold ventures taking place in the UK, to assist with the monitoring of ventures abroad and to collect statistical information which will facilitate ventures in the future.

The Award has a number of Local Guides relating to specific designated Wild Country Areas in Europe which are available, on request, from the European Department based at the Award's South East Regional Office. The range includes the Cevennes, the Ardennes, Los Picos de Europa, the Pyrenees, the Harz Mountains and the Chamonix area.

The Youth Exchange Council and the Commonwealth Youth Exchange Council both provide general information on exchange visits (see Appendix).

Joe Cornish

Open Golds

Open Golds provide an opportunity for independent, unattached, or other participants at Gold level who are unable to form a viable group, to carry out their qualifying venture in order to complete the Expeditions Section. Open Golds are organised by many Operating Authorities and Wild Country Panels and last for six or seven days which includes familiarisation, team building and planning periods, followed by a supervised and assessed qualifying venture.

Many Open Golds are advertised in *Award Journal* but a telephone call to the Territorial or Regional Offices will usually help participants to find out what other opportunities are available. They usually take place at Outdoor Education or Pursuits Centres which are located in designated Wild Country Areas and are staffed by qualified Centre staff, members of the local Wild Country Panel or Supervisors and Accredited Assessors from an Operating Authority. Open Golds usually prove to be an exciting and challenging climax to the Expeditions Section.

As Open Golds last for only six or seven days it is necessary for all participants wishing to take advantage of these opportunities to be fully prepared, trained and equipped and with the required number of practice journeys completed. Participants are expected to arrive with the necessary levels of physical fitness for an Expedition. There is only time in the two or three days prior to the venture to form groups, check training and equipment, plan the venture and prepare the tracings and route cards. Leaders running some Outdoor Centres have items of camping equipment which may be borrowed, but this should be checked in advance.

The vast majority of Open Golds offered to participants are Expeditions involving 80 km (50 miles) of journeying on foot through wild country. These are, undoubtedly, the preferred option of most participants. The Award would like to see a greater number of alternative forms of venture, involving other modes of travel, but it recognises the difficulties in recruiting sufficient numbers of participants with the necessary skills and training related to the mode of travel, such as canoeing, sailing, cycling or horse riding to guarantee a viable venture. Explorations may prove difficult to arrange because of the degree of prior preparation necessary and the often diverging interests of the participants.

ADVICE FOR PARTICIPANTS

The following advice is offered to participants wishing to avail themselves of the opportunities offered by Open Golds.

Those who seek Open Golds tend to be participants who are in employment or who have left the Unit to which they were attached for college or university, though anyone over the minimum age for a Gold level venture would be considered. The majority are 18 years of age or over, which is of great help in overcoming many of the problems with which they are faced. Access to the *Expedition Guide* and the *Gold Entrance Pack* is necessary for participants wishing to take part in an Open Gold as they must know the requirements and conditions of the Expeditions Section which apply to all Gold qualifying ventures.

Participants must have:
- The necessary clothing and equipment.
- Completed the training syllabus for Gold level.
- Completed the necessary number of practice journeys.
- Had the appropriate page in their *Record Book* signed to confirm that all the training has been completed.

Advice on clothing and equipment is detailed elsewhere in this *Expedition Guide,* as is the training syllabus. The *Programmes File* and Award's website provide an itemised emergency equipment list as well as the complete training syllabus. The first aid training is likely to prove difficult for many isolated participants to complete. If a participant has moved to a different part of the country, the appropriate

Joe Cornish

Award Secretary or Regional Officer may be able to put the participant in touch with the nearest Award Unit where assistance may be sought.

It may be possible for independent participants to join a local Unit for practice journeys; otherwise they will have to devise their own journeys accompanied by friends (see Chapter 5.2 - Effective Practice Journeys). Every journey must be supervised by an adult approved by the participant's Operating Authority.

Participants must acquire the necessary level of physical fitness before attending Open Golds as there will not be time for this to be achieved during the two or three days' preparatory period. Fitness training will be most effective if it is commenced two or three months before the Open Gold takes place and becomes progressively more strenuous.

At the outset, thought should be given to the purpose of the venture. This may be one which participants can complete as individuals without involving others; though if, when the groups have formed, others can become involved, so much the better.

Finally the organiser should be informed of any medical condition which might affect performance or supervisory needs and, if catering is provided, of any special dietary needs. This is usually covered by the organiser sending out the Medical and Consent Forms.

ADVICE FOR FOR THOSE ORGANISING OPEN GOLDS

Open Golds require careful timing and only experience or consultation with those who have experience in organising these events can indicate the most suitable dates for them to take place. School holidays or Bank Holiday weekends plus the remainder of the week during the Expedition season are probably the most appropriate. The timing should meet the needs of the participants rather than those of the Centre. It is helpful if there is a spread of different weeks over the country as a whole to provide a wider range of opportunities for participants.

Open Golds should be advertised well in advance if they are to achieve viable numbers. Advertisements for inclusion in the *Award Journal* should reach the Communications Department at Award Headquarters by the end of October each year to be included in the Spring issue which is published in the first week of January. There are many other avenues for advertising through Operating Authorities and individual Units. The Territorial and Regional Offices of the Award Scheme should always be informed unless they were involved in setting up the Open Gold. The cost should be indicated and include any 'hidden extras' such as the hiring of equipment. The advertisement should state where further details may be obtained.

A full information pack and application form should be prepared at the same time as the advertisement. Full and detailed information will reduce future correspondence with the participants who will be better prepared and feel more secure.

The following information should always be included:
- The form of venture, or ventures, on offer and a brief description of the terrain or water involved.
- Equipment which is provided, or which may be borrowed. If maps are not provided then the reference numbers of the maps required should be included.
- Details of the catering arrangements should state whether food will be provided before and during the venture or whether participants will have to cater for themselves. If food is provided, the ability to cater for special dietary needs should be mentioned.

The following information should always be sought from the applicants:
- Any medical condition which may/will require special attention.
- Any dietary needs or restrictions.
- The time when the participant hopes to complete training and practice journeys and who will sign the appropriate page of the *Record Book.*
- Whether the form of the presentation has been considered and to whom it will be submitted.

Organisers must ensure that adequate insurance cover is in place for all Open Gold ventures.

RESPONSIBILITIES

The organiser and staff of Open Golds have a duty to carry out the role of responsible Leader or Supervisor to the participants, so that they may enjoy the same standard of care and advice in their preparation and planning as they would if they were members of a normal Award Unit. They are disadvantaged when compared with participants who have enjoyed regular training with a group of companions and they will usually require extra advice and attention.

Participants, where possible, should choose their own companions and form their own groups. Experience has shown that arbitrarily formed groups frequently experience stress and incompatibility which have jeopardised the opportunity for success in the past. There are, however, considerable advantages in participants with similar levels of physical fitness venturing together. Friends should have the opportunity to stay together. Some organisers group participants into geographical areas and send out lists of the those attending with addresses and telephone numbers, and encourage the participants to share transport and equipment and start to plan their routes together. After the groups have been formed, the appropriate Wild Country Panel Secretary should be contacted to obtain a notification number to be entered in the participants' *Record Books* following the successful completion of the venture.

Every assistance should be offered to facilitate the formation of groups and the integration of individuals into effective teams. This is best achieved through the planning and preparation for the venture, coupled with regular reviewing sessions.

Shortly after arrival, participants should receive a Pre-expedition Check of equipment and training so that they and the staff have an opportunity to remedy any deficiencies which are apparent. Weighing scales should be available on arrival so that candidates may correct overweight packs well in advance of the venture.

Staff should use every available opportunity to improve the competence and performance of the participants during the two or three days' preparation period. The Pre-expedition Check will help in diagnosing where assistance is required. Attention must be given to the purpose of the venture and the subsequent report. The alternative forms and methods of presentation should be discussed and the participants informed that the choice is entirely theirs.

Where participants are geographically isolated and have no local support, it may be advantageous to encourage them to submit an oral report after they have been debriefed and had time to prepare their submission. This ensures that the *Record Book* is fully signed to denote the successful completion of the Expeditions Section.

Adequate time should be set aside during the day for the newly formed groups to plan their routes, alternative and escape routes and produce route cards and tracings. Route planning and the production of route cards is very time-consuming and sufficient time must be allowed for this process. This should not take place on the evening before the venture and interfere with the opportunity for a good night's rest.

Provision should be made for the participants to receive current weather forecasts.

The most useful practices are short journeys or sessions, which explore the participants' competence in aspects such as navigation, camp craft and emergency procedures which should be followed by frequent reviewing and individual or group tuition.

Care must be taken to ensure that the participants are not tired or exhausted before commencing their qualifying venture. They are entitled to set out, after a good night's rest, carrying dry equipment in dry packs and with their feet in good condition.

For the period of the venture the group should have a nominated Supervisor who should perform the duties and daily visits outlined in Chapter 6.2 - Supervision. Assessors should be appointed who will co-ordinate their visits with those made by the Supervisor.

After the completion of the qualifying venture and when the participants have had time to clean up, the usual debriefing by the Assessor should take place in the presence of the Supervisor. Participants wishing to submit oral reports should make their report after ample preparation time. Groups which disperse immediately after they are back at base derive less benefit from the total experience.

Experience has shown that despite the difficulties of training, preparation, planning and journeying with their group for the first time, most participants find that the successful completion of an Open Gold Expedition clearly meets the aims of the Expeditions Section in developing a spirit of adventure and discovery.

EXPEDITION GUIDE

3

THE DUKE OF EDINBURGH'S
AWARD

Part

3

ADVICE
AND SKILLS
COMMON
TO ALL
VENTURES

Part 3
Advice and Skills Common to all Ventures

Equipment

T he careful selection of equipment is not only important for safety and well-being, but vital to the comfort and enjoyment of any venture. This means not buying the most expensive, but seeking the advice of those experienced in the outdoors and making the right choices.

Many Operating Authorities have pools of camping equipment, sometimes even waterproof overclothing, which may be borrowed or hired. It is a good idea to take advantage of this if you can, as it gives you the opportunity to test and gain experience of a variety of equipment so that you can make informed choices when purchasing your own. All participants should have their own clothing, waterproof overclothing and personal and emergency equipment; it is desirable that they should have their own rucksacks and sleeping bags. The Award Scheme is concerned with extending the range of personal interests and it is hoped that the Gold qualifying venture will lead to many more outdoor ventures rather than be the last.

All participants should make their own checklists of equipment when they start to prepare for their first practice journey. This can be based on the lists in this *Guide,* with items being added or deleted according to personal needs and experience. The list should be on stout card and carefully preserved so that it can be used to check equipment before departure on future ventures. It is all too easy to forget just one item which may turn out to be of vital importance. It is convenient to divide equipment into the following five categories and lists can be made under the following headings:

- Clothing
- Personal and emergency equipment
- Personal camping equipment
- Group camping equipment
- Equipment related to the mode of travel

Equipment related to the mode of travel will be found in the relevant chapter of Part 4 - Advice and Skills Associated with the Mode of Travel.

CLOTHING

Most items of clothing are the same for all modes of travel and the principles of dressing for the outdoors are always the same. Any variations are dealt with in the chapters concerned with the mode of travel.

Clothing must be capable of protecting the wearer under the worst conditions which may be encountered, and the cold, wet, windy conditions which can be experienced at any time of the year in the British Isles provide the severest of tests. The rapidly changing and unpredictable conditions make the wearing of suitable clothing all the more difficult.

Insulation is provided by the air trapped between the fibres of cloth and between the layers of garments, so two light pullovers weighing around 450 grams each provide more insulation than one thick pullover weighing 900 grams, and give the added advantage that you can add or remove them to regulate your temperature more easily. This is the basic principle in dressing for the outdoors - layers of clothing which trap air and which can be removed or added to as the weather and the amount of physical exertion demand.

Garments should be loose fitting to trap air between the layers, to provide insulation and to allow air to circulate in hot weather. Outer garments must therefore be adjustable. Zips are essential; they enable outer garments to be opened or closed at the front. Fastenings at the neck, cuffs, waist and below the hips enable the wearer to increase or reduce air circulation.

Clothing loses most of its insulating qualities when wet, whether from rain or from perspiration, so it essential to keep clothing as dry as possible. This entails reducing sweating when working hard by opening or removing clothing, and covering normal clothing with waterproof overclothing when it is raining. When waterproof overclothing is worn, perspiration cannot escape so the clothing becomes damp, but at least the perspiration is warm and heat loss is reduced. Modern breathable fabrics go a long way to reducing the soaking of clothing by perspiration.

There are several materials which retain most of their insulating properties when wet: the traditional natural fibre wool, and modern synthetic fibres made up into fibre-pile garments. Modern synthetic fibre garments offer very good insulation at very light weight. They have the advantage over wool in that they do not absorb as much water and do not increase in weight to the same extent. Like wool, synthetic fibre garments need great care in washing and drying if they are to keep their size and shape. Washing and drying instructions should be followed with meticulous care. If wool is used, and many still use woollen sweaters, a mixture of 60%-70% wool with a synthetic fibre is usually more suitable for outdoor needs.

Clothing is very much a matter of individual choice and experienced backpackers differ widely in their preferences. The majority find that modern synthetics, thermal underwear and fibre-pile shirts and jackets suit their needs, while some traditionalists still favour wool and cotton - 'cotton cools - wool warms'.

Climbers and walkers talk of 'inner layer', 'middle layer' and 'outer layer', or 'shell'. Regardless of the fabrics used, it is customary to have an inner layer to absorb perspiration, a middle layer to provide insulation and an outer layer, or shell, to keep the wind and rain out.

The advice outlined above is aimed to give protection in the wet, cold conditions which can occur at any time during the Spring, Summer and Autumn in the British Isles, but it must always be remembered that more ventures are aborted in hot weather than in cold, and equal attention must be given to protection from the sun and heat. Many more ventures are now taking place outside the United Kingdom and clothing must be able to cope with the temperatures on the Continent. Loose fitting cotton clothing is most people's preferred choice for hot conditions.

Inner Layer Clothing

Modern synthetic fibre underwear is very popular and in colder weather thermal underwear is a great help, especially in water ventures, but again it does require careful washing and drying. In recent years there has been a tendency to wear closer fitting garments utilising modern stretch fabrics which do not impede movement to the same extent as tight fitting garments in natural fibres. These synthetics have excellent wicking qualities to remove moisture from the skin.

Do not neglect to protect the lower part of the body against cold, wet conditions; there is a wide range of thermal garments available, such as long johns and leggings, which absorb very little moisture and dry very quickly. For many, cotton underwear still has great merit as it is pleasant to wear, absorbs perspiration, dries quickly, is relatively inexpensive and, because it requires little care in washing and drying, stands up to the hard wear of Expeditions.

Middle Layer Clothing

Insulating fibre-pile, fleecy garments in modern synthetic materials are most people's choice in outdoor pursuits and are often worn as outer garments in dry weather. They are lightweight, windproof and dry very quickly. Because they absorb so little moisture they remain light, even when soaked. Others still prefer the long flannelette/brushed cotton working shirts which are cheap and stand up very well to hard wear. Whatever kind of garment is worn, there should be no gap at the waist as is so often the case with many T-shirts and sweatshirts worn today. A pair of light or medium weight, loose fitting woollen pullovers with long sleeves provide a cheap and most effective method of keeping warm. In hot conditions a loose fitting cotton shirt with long sleeves and long cotton trousers are cool and provide good protection against the sun.

Synthetic or breathable fabrics have virtually replaced the worsted, cord, twill and tweed trousers. Traditionally, trousers used for outdoor activities tended to be loose and baggy to permit freedom of movement and to trap air, but the cut of garments in modern synthetic stretch fabrics is much closer. Jeans are unsuitable as they are often cut too tight, become very heavy and provide little protection when wet, and take a very long time to dry.

Overclothing should be removed in warm weather.

Outer Layer Clothing

The purpose of the outer layer is to provide protection against the chilling effect of the wind and rain, and it essential that garments fulfil this function. Even the most 'high-tech' breathable fabrics have difficulty in coping with sweat from working hard in severe weather conditions, and the clothing underneath may become wet with perspiration; it is important that the wearer is able to exert the maximum control over ventilation. The waterproof overclothing which forms the outer shell may be made from a breathable fabric, or from a totally waterproof fabric, incorporating a neoprene coating. A jacket with a heavy-duty double-ended zip is always preferable to a smock as it enables the garment to be opened for increased ventilation and ensures free leg movement. A wired hood with draw-cord and large enough to fit over a hat, adjustable storm cuffs and a draw-cord round the hips or waist will provide maximum control over air circulation. If the clothing underneath is not to become soaked in sweat and lose its insulating qualities, waterproof jackets should not be worn all the time but should carried on top of the pack and only worn when it is raining .

Modern breathable overclothing must be kept very clean if it is to work efficiently. If the micropores are clogged with dirt then the garment ceases to function effectively. The maker's washing instructions should be followed, or garments should be washed with a pure soap or one of proprietary cleaning agents.

Some people may prefer to use heavy-duty (6-8oz or 150-200 grams) neoprene or polyurethane proofed nylon garments with taped seams which are much cheaper than garments made from breathable fabrics and are completely waterproof. They, and the wearer, tend to become soaked in perspiration but they are very hard-wearing and stand up well to the chaffing of backpacks and the hard use and abuse on Expeditions in the wet conditions of the British Isles. Outdoor workers in all trades and occupations use these garments, and many outdoor centres prefer them for their hard-wearing qualities and relatively low cost.

Waterproof overtrousers usually complete the outer protective shell, but some walkers still prefer to use a long cagoule reaching below the hips in conjunction with long johns, leggings or even a pair of track suit bottoms under their walking trousers in cold conditions.

Waterproof overtrousers have to be put on over boots so they must be wide enough in the leg or have zips at the ankles.

Socks

Socks perform three functions: they cushion the feet, absorb perspiration and provide insulation against the cold. The majority of walkers seem to prefer two pairs, a combination of short and long, or thick and thin. A mixture of wool and man-made fibres is recommended, and they should be free from holes or darns. Frequent washing is necessary for them to function properly and at least one spare pair, and preferably several, should be carried. In hot weather wearing two pairs of socks will probably soften the feet and increase the risk of blisters. Instead of boots, one pair of socks worn with a suitable pair of trainers will help to reduce this risk if the conditions underfoot permit.

Headwear

As an unduly large proportion of the body's heat is lost through the head, it is essential to wear some form of protection if you are to remain warm. The woollen balaclava is a popular and most effective way of protecting the head and represents very good value for money, but home-knitted woollen hats are quite suitable providing they are large enough to come down over the ears. There is some very effective, lightweight synthetic fibre headwear available on the market. In heat and strong sun, a light, wide-brimmed hat provides protection, not only for the head, but for the face and neck as well.

Gloves

There is a wide range of suitable gloves available for outdoor use for all the various modes of travel. Woollen mittens or gloves are a cheap and effective method of keeping the hands warm as the wool quickly matts and becomes windproof and the hands stay warm, even when the gloves are wet. All gloves should be sufficiently long to protect the wrists and tuck inside jacket sleeves.

The clothing listed above is necessary for anyone undertaking a venture in the British Isles, even in the middle of summer, because of the high rainfall and the unusually high winds in the upland areas. Ventures on the Continent, especially the South, or in other hot climates, may necessitate different clothing - loose fitting cotton, or cotton/synthetic mixture for long-sleeved shirts and trousers to protect the arms and legs from the sun. Trousers are preferable to shorts as they protect the legs from the sun, as well as giving much greater protection against insect bites, scratches and abrasions.

Boots

Of all the clothing and equipment used, nothing will have a greater impact on the enjoyment of any walking venture than the footwear. Boots should be light - saving 500 grams of weight in the boots is equivalent to saving several times this amount in your rucksack -

remember that your feet have to travel twice as far as you do on a journey! Boots have improved greatly in design as the research and development of trainers has been extended to walking boots. They are lighter, much more comfortable and, incidentially, cause less environmental damage. The most important function of a walking boot is to prevent you from slipping or falling - the most common type of accident needing the services of mountain rescue teams are those involving breaks or sprains of the lower limbs. A flexible cleated rubber 'Vibram' type sole is most suitable. The sole should be thick enough, and the boot sufficiently well padded, to absorb the pounding which your feet will take.

Traditional hill and mountain walking boots are made with a one piece leather upper with the smooth surface on the outside and a bellows tongue, which enables the boot to be immersed in water up to the ankle without getting the feet wet. D-rings and hooks to make it easier to put them on and take them off. Such boots are easy to clean, maintain and protect with wax, silicon and boot oils. There is a wide range of boots made in this traditional form on the market but they tend to be more expensive. There are also boots made of suede and fabric which are lightweight and reasonably priced.

When purchasing boots always take with you the socks which are going to be used so that they may be worn when boots are tried on. Choose carefully and take your time, as you may spend a great deal of time regretting the choice of ill-fitting boots.

Modern lightweight boots do not need as much breaking in as the old-fashioned heavier ones, but they do need to be broken in and this can take several weeks. Wear them around the house, to work, college or school. To embark on a venture in boots which have not been broken in is a recipe for disaster. For traditional leather boots, liberal applications of boot oil or liquid dubbing will help the process, while some walkers prefer to give them a really good soaking in water, walk in them until they are dry and then apply the boot oil. Once broken in, a regular application of ordinary wax polish will keep them supple and waterproof. Suede and fabric boots do not take long to mould to the foot. The manufacturer's instructions concerning their preservation must be followed. Borrowing boots which have been moulded to someone else's feet is borrowing trouble!

A pair of trainers should be carried for use around the camp site and, provided they have robust non-slip heavily cleated or studded soles, they are excellent for walking in hot dry weather. Trainers are not as comfortable on stony or steep terrain, they do not provide anything like the protection against the wet and cold, and are not suitable with gaiters.

Gaiters

Gaiters provide the link between footwear and the rest of the clothing, helping to keep the feet warm and dry in bad weather and when conditions are soggy underfoot. There must be zips at the side or back to enable them to be put on or taken off without removing the boots.

PERSONAL AND EMERGENCY EQUIPMENT

In addition to the clothing listed above which will be worn or carried in the rucksack, there is a small number of items of emergency equipment which should always be carried by every member of a group in wild country. Personal and emergency equipment is virtually the same for all modes of travel, but some variations do occur especially in water ventures. The equipment should be kept to the bare minimum or the exercise in safety will become self-defeating with extra weight leading to fatigue and a loss of mobility.

Spare Clothing

This may range from a spare sweater, socks and scarf for a day journey, to complete changes of clothing for a camping expedition.

Bivvy Bag

A heavy-duty polythene bag, large enough to get inside, is a proven life saver. Gram-for-gram there is probably no item of emergency equipment which will do more to assist survival in adverse conditions than a bivvy bag. Along with spare clothing, it must rank as one of the most important items of emergency equipment for each individual. When a member of a group is suffering from incipient hypothermia, it is probable that other members of the group may be approaching the same state; it will not help if they have to queue up to take turns in a couple of bivvy bags.

Emergency Rations

Chocolate bars, nuts and dried fruits, boiled sweets or mint cake provide effective and easily obtained iron rations.

Whistle

A plastic whistle should be carried, either in a pocket or, better still, on a lanyard round the neck.

Torch

Each member of the group should carry a medium-sized torch fitted with alkaline cells or batteries, together with a spare bulb and spare alkaline batteries. Alkaline batteries last five or six times longer than the old zinc/carbon types. They have a better shelf-life, so the chances of buying time-expired ones are greatly reduced, and are much more reliable in emergency conditions than the rechargeable nickel-cadmium (Nicads) or nickel metal-hydrides. Remember that it is impossible to read the map if you can't see it! There is no need to carry a heavy searchlight around - the need is for a modest amount of light over a prolonged period of time, especially in the Spring and Autumn when the days are shorter and the use of torches around the camp site is increased. A headtorch is a useful addition to any equipment list as it leaves the hands free.

Map(s) and Compass

Each individual member of the group should have the appropriate map or maps of the area and a compass. The maps do not have to be to the same scale, some members will prefer to carry the 1:25 000 while others will carry the 1:50 000 scale. A map case will provide the necessary protection.

Pocket Knife

A simple high quality pocket knife which will retain a sharp edge is better than a poor quality knife with a dozen different gadgets. A tin opener and a spike are the most useful additions to the blade.

Matches

In addition to one or more boxes of safety matches carried in the driest place you can find, spare emergency matches should always be carried. The matches and the sandpaper from a box of matches should be placed in a plastic 35mm film container for protection.

Pencil and Notebook

These will be needed by each individual and can be carried in the map case or pocket for easy access. Pencils are more reliable than pens in wet or adverse conditions, especially if they are sharpened at both ends.

Water Bottle

There is a long tradition of climbers and hill walkers not drinking while on the hill, or drinking from mountain streams. It is questionable whether this tradition can be supported within the Award Scheme, especially in hot conditions in the areas where most ventures take place and where stream water may be, at best, of dubious quality. A water container may be regarded as essential personal equipment. See Chapter 3.3 - Catering for Expeditions.

The equipment listed above will go a long way to ensure the safety of all the individuals within a group in the worst conditions likely to be encountered during the Expedition season. The desire to carry additional items should be resisted. Items such as ice axes are inappropriate for Award ventures and, in inexperienced hands, only add the dangers of disembowelment to the other wild country hazards! Each individual should carry all these items. Even with the best of intentions individuals, on very rare occasions, will become separated from the rest of the group, especially if help is being sought. All groups must be self-reliant and each individual within the group must be able to survive independently of the rest of the group.

Where all members of a group have all the items of emergency equipment, it provides a reservoir of equipment which will overcome the loss or malfunction of individual items. While spare clothing, bivvy bag, whistle, emergency rations and the torch are a must for each individual, a group should never be accused of being inadequately equipped if every member of the group is not in possession of a map, a compass, pocket knife, pencil and notebook, or spare batteries and bulb. It is essential that a group carries at least two maps and compasses.

PERSONAL CAMPING EQUIPMENT

Personal camping equipment is similar for all modes of travel with the exception of the rucksack. The alternative provision is covered in the appropriate chapters.

Rucksacks

One factor which distinguishes foot ventures from all the other modes of travel is that all food and equipment is to be carried on your back and the choice of pack is therefore very important. Although a rucksack is very much a matter of individual choice and fashions change, those with an

BERGHAUS NITRO RUCSAC

Berghaus' Nitro rucsac is ideal for fast moving, exhilarating sports such as rock climbing, mountain biking and trail running.

Combining the latest in fabric technology and state of the art construction, the Nitro features the 'Limpet' integrated compression system which moulds the sac against the body, increasing stability. This ensures the rucsac remains stable during energetic outdoor activities and won't hinder or impair sporting performance. The Nitro was awarded a Millennium Product Award for design and innovation.

Nitro has a 24 litre capacity, providing enough space to stow away all necessary essentials. It also includes a mesh helmet net, extended from the base pocket to provide extra capacity.

1. The shield is connected directly to the hipbelt and shoulder harness to enable the load to be pulled tight to the body for maximum stability.

2. Super smooth laminated harness construction spreads the load more evenly for greater comfort

3. Kinetic 3D mesh keeps the back cool and absorbs no moisture.

A: Firstly, tighten straps around hipbelt

B: Secondly, tighten straps over shoulder harness

BERGHAUS FREEFLOW RUCSAC

The Freeflow daysac is part of Berghaus' Trekking range. It features the innovative Freeflow air circulation back system, which is composed of a pre-stressed ABS board with a mesh back that together keep the sac away from the body and allow for the easy movement of air. The Freeflow system is ideal for walkers who want to avoid 'sweaty back sydrome' and remain comfortable when out on the hills.

The Freeflow range is available in capacities of 18,23,25,28,30+6,35 & 35+8 litres. Other features on the Freeflow include: side pockets (both big enough to hold a one litre water bottle), external lid pocket, internal key security clip, ice axe/walking stick holder, contoured shoulder harness and padded hipbelt.

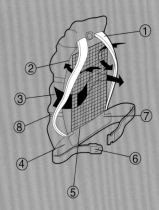

1. Contoured shoulder harness for stable and comfortable fit

2. Tough ABS stiffener doesn't collapse under load

3. Large open mesh gives maximum contact with air

4. Fully padded hipbelt

5. Non-absorbent mesh (ie doesn't soak up perspiration)

6. 35mm waistbelt

7. ABS board pre-stressed in three directions for maximum airflow

8. Available in two back lengths - standard (18, 23, 25, 30) and long (28, 35, 30+6, 35+8)

integral frame dominate the market. The rucksack should be large enough to contain all your equipment and, on other occasions, be suitable for holiday, travelling and social needs. A rucksack is usually the first investment that a young person interested in the outdoors makes in the way of equipment.

A capacity of 65 litres should meet all these needs, though young women may prefer a 55 litre capacity pack. Tough 200 gr (7-8 oz) polyurethane proofed nylon is a suitably robust material which will stand up to the wear and tear of everyday use. The pack must be tested for fit and comfort. The shoulder-straps should be wide and well padded. A well-padded hip belt is essential in ensuring a better distribution of the load, taking weight off the shoulders and increasing security on steep ground.

Sleeping Bags

It is very difficult for one sleeping bag to cope with all the extremes of temperature which can occur throughout the year. What is known as a 2-3 season bag, with a synthetic hollow fibre filling, should meet all the needs of those involved in the Award Scheme. It is bulkier than a down/feather bag, but it is much cheaper, provides good insulation when wet and can be washed with care. One of the various forms of over-lapping or box quilting will ensure good insulation and avoid cold spots.

Inner Sheet Bags

An inner sheet bag should always be used with borrowed sleeping bags. Individuals fortunate enough to own a down or feather bag should regard a lining as essential to keep their bag clean and avoid the necessity of washing, but those who possess their own synthetic filled bags usually prefer to do without a liner and wash the bag occasionally.

Sleeping Mats

Backpackers of today regard closed-cell foam sleeping mats as essential. They go a long way in ensuring a more comfortable night's sleep and are very effective in increasing insulation against the cold from the ground, as well as being a very useful item of emergency equipment. As the foam does not absorb water they can be strapped to the outside of the rucksack.

Changes of Clothing

In addition to the spare emergency clothing listed elsewhere, Award participants should carry a complete change of clothing for ventures lasting more than a day. This will ensure that they will always be able to change into a set of dry clothing. It is customary for those engaged in outdoor activities to put on wet clothing in the morning if necessary, in an effort to preserve one set of dry clothing in case of an emergency. A track suit provides ideal night attire and is useful around the camp site. They are also less conspicuous than pyjamas when they have to be worn in inhabited areas when all else is wet.

Eating Utensils

A plastic mug, a plastic or aluminium plate or mess tin and a set of cutlery is all that is required by each individual; everything else can be shared.

Washing Gear

Soap and towel, toothbrush, toothpaste and toilet paper or 'wet wipes' completes the usual list of personal equipment needed by the backpacker when travelling as a member of a group.

GROUP CAMPING EQUIPMENT

Tents

There are dozens of lightweight tents available for campers to choose from and they come in all shapes and sizes. Good quality tents are never cheap so, if it is possible to borrow a tent, do so. Many Operating Authorities have tents which they will lend or hire out and about 80% of Award participants succeed in borrowing a tent, enabling them to give priority to purchasing the more personal items such as the rucksack or sleeping bag. Before buying a tent, it is useful to try and borrow one of the same kind so that you can examine the quality of construction and try it out for size and convenience. Lightweight tents usually hold two or three people and the load can be shared between the occupants. Tents usually have an inner tent with a sewn-in tray ground sheet and an outer cover. The traditional ridge tent still provides good value for money, but with the large variety of tents on the market in different fabrics it is impossible to make specific recommendations.

The pegs provided with a tent are not necessarily most suited to your needs, and it is usually worthwhile replacing some of the pegs with others of a different pattern so that you can anchor your tent securely in different terrains. If a borrowed tent is to be used, it is essential that it is pitched before setting off on the venture to ensure that it is complete and that you know how to pitch it, even in the dark.

Margaret Dixon

Cooking Stoves

Award participants should consider their future needs when purchasing a stove. A stove which is looked after will give many years of trouble-free service and may well last a lifetime. Weight, bulk and efficiency are the most important considerations, as well as the availability of fuel. The mode of travel plays little part in choosing an appropriate stove as it will always be transported by your own physical effort. As with tents, if you are able to use a borrowed stove, take advantage of the opportunity as it will help to guide your choice in any future purchase.

The most popular choices are stoves fuelled by methylated spirits and gas. Spirit stoves, serving a duel function of stove and canteen, dominate the market, but gas stoves of improved design and using self-sealing cartridges are becoming increasingly popular. **Never purchase a gas stove which does not use self-sealing cylinders or cartridges**.

To stress the need for safety when using cooking stoves, Chapter 3.4 has been devoted to providing advice on choosing a stove and giving advice on their use.

Fuel Bottles

Fuel should be carried in bottles with a secure screw top which have been specially designed for the purpose. There are many excellent bottles available of suitable capacities.

Canteen/Cooking Set

A suitably sized canteen, or a set of pans, will be needed to cater for each tent group unless a combination spirit stove is being used. Canteens are of the nesting type where pans, usually three in all, fit inside each other with the lid serving as a frying pan.

Water Container

A collapsible plastic water container is essential to store water for each tent group.

Washing-up Materials for the Group

Wire-wool soap pads are a firm favourite for dealing with greasy or sooty pans. A small quantity of liquid detergent in a small plastic container, a nylon scouring pad and dishcloth or sponge will take care of the rest of the utensils and dishes. Part of a well-used tea towel, a piece of cotton rag, or a dish-cloth will be needed for drying utensils and for wiping up any spillage in the tent.

A trowel and a tin opener will complete the group's camping gear, but a needle and thread as well as a hank of thin cord or guy line will be worth their weight in gold.

Finally, it is worth remembering that adding more equipment to the list rarely increases safety - it just increases the probability of succumbing to fatigue and exhaustion, and reduces mobility. When faced with difficulties or an emergency it is most important that groups make the most effective use of the equipment and resources they have available, or in the environment of the immediate vicinity. Pooling, sharing, and above all else, improvising, provides the solution to most difficulties. With improvisation, or divergent thinking, most items in your pack will serve purposes other than those for which they were intended. A triangular bandage will serve as a sling, a flag, a cold compress or protection for the head or neck, to mention just a few roles. Practically all other items of emergency equipment, camping gear or clothing can be utilised to provide solutions in an emergency.

Camp Craft

amp craft could be defined as the ability to provide oneself with food and shelter under all the conditions likely to be encountered in the outdoors. The ability to do this improves greatly with practice but the effort is worthwhile, not only for its use within the Award Scheme but in later life, for the satisfaction and enjoyment it can bring in its own right and for recreation and social activity. Camping also provides a freedom and independence to participate in a whole range of other outdoor activities. **The care and protection of the environment in which the activity takes place must always be at the centre of the activity**. Camp craft divides conveniently into two parts:

- The provision of shelter
- The provision of food

The Award Scheme is always concerned with high standards and these standards should be as apparent in camp craft as within any aspect of the Award programme. Camping can easily degenerate into living rough or, as the French would say, 'camping au sauvage', and it is essential that the highest standards are adhered to at all times. There is no need to stress the importance of camp craft for our well-being and safety, for the ability to provide ourselves with food and shelter is a skill on which our very survival may depend. The excellent safety record of the Expeditions Section is largely based on the insistence on camping as the means of accommodation, with all members of the group having their camping gear and food to fall back on, regardless of the nature of their venture or the mode of travel being used. See Chapter 3.3 - Catering for Expeditions for the provision of food.

EQUIPMENT

As stated in the previous chapter, camping equipment divides into personal camping equipment, and group camping equipment to be shared between the occupants of a tent. Two or three people should be considered as the **basic unit** occupying each tent, with two or more tents forming the **Expedition group**. It is possible of course, especially at Bronze level, for the minimum number of four to be accommodated in one tent. The occupants of a tent, working on the 'buddy system', will share the activities and workload between them. The stoves and the other equipment related to catering will be considered in the next chapter.

WATERPROOFING, LOAD CARRYING AND PACKING

Waterproofing

Successful camping begins at home with the waterproofing of certain items of equipment and the packing of gear into the container(s) in which it will be carried. The containers will range from the rucksacks to the panniers for cycle ventures or waterproof containers used by those engaged in water ventures. The details of the waterproofing process are outlined in the relevant chapters dealing with the mode of travel. The policy should be to make the rucksack or pannier as water resistant as possible by using a strong waterproof plastic liner, and then individually waterproofing items such as the sleeping bag, spare clothing and food in smaller plastic bags tied with elastic bands; these do not pierce the bags as do the metal/plastic ties and spare elastic bands should be included. Bin-bags rip and tear too easily, so a cheap solution is to obtain an empty fertiliser bag from a farmer or gardener, or a similar bag from a builder's yard, as they are usually made of 1000 gauge polythene.

Load Carrying

Irrespective of the mode of travel, the weight of the load must be kept to a minimum. For those venturing on foot the load should not exceed one quarter of the individual's own bodyweight. All packs, panniers and waterproof containers should be weighed - bathroom scales are ideal for carrying out this procedure. The lighter the pack the greater the enjoyment! After the basic camping and safety equipment has been assembled, additional items will not increase the margins of safety but

Sandra Skinner

*No matter how expensive, rucksacks are never waterproof -
they always need a heavy poly-bag liner*

will only slow you down and add to the discomfort and dangers of exhaustion and, in foul weather, increase the possibility of hypothermia or, in hot weather, the probability of exercise-induced heat exhaustion.

Considerable self-discipline is needed to keep the load to a minimum. The Latin word for baggage/equipment is 'impedimenta'! Inexperienced campers are often inconsistent in their attitude to weight. They will go to endless trouble to find the lightest tent and even count the number of tea bags, only to waste all their efforts by carrying a single-lens reflex camera with a telescopic lens weighing over a kilo, a giant lump of soap and a towel large enough to dry an elephant! All unnecessary items must be eliminated; it can never be more than a sliver of soap and a small well-used hand towel. Many of the problems arise from packing at the last moment and then throwing in whatever comes to hand. **Equipment should be assembled well in advance** for, like route planning, it always takes longer than you think.

Packing

All gear should be carried within the pack, with the possible exception of the foam sleeping mat. If tent poles are carried externally, great care should be taken to ensure that they are properly secured in a bag. Another exception is clothing which is being dried while you travel. All wet or damp clothing and socks which have been washed should be dried as soon as possible. This will not only add to your future well-being but lighten your load. Clothing attached to the outside of the pack must be firmly secured and checked at regular intervals

THE CAMP SITE

Finding a suitable camp site is not only important for comfort but may affect one's safety and well-being. The criteria used for camp sites for Award Expeditions may differ considerably from holiday camping, but care taken in choosing the right location will be amply repaid in comfort and freedom from unexpected and unwanted happenings during the night or the following morning.

A camp site suitable for mobile lightweight camping must fulfil a number of criteria. It should be:

- Able to provide shelter from the wind.
- Permissible to camp on the site.
- Free from objective dangers.
- Able to offer some degree of privacy.
- Within easy reach of water.
- Reasonably level.
- Able to provide a secure anchorage for the tent and sufficiently soft to offer the prospect of a good night's sleep.

Able to provide shelter from the wind - Award participants with lightweight tents will be concerned with finding shelter from the wind, which will not only significantly reduce the possibility of the tents being flattened, but will facilitate cooking and make life around the camp site more enjoyable. This usually means seeking lower ground, hollows or the lee side of a ridge, hill, wood, hedge or wall. However, in Spring and Autumn under clear skies, cold air can sink down slopes and collect in hollows, the so-called 'frost hollows'. If there is little or no wind when the tents are pitched, then it is usual to take shelter from the prevailing wind which in the British Isles is from a Westerly direction. Place the windbreak between the tent and the wind.

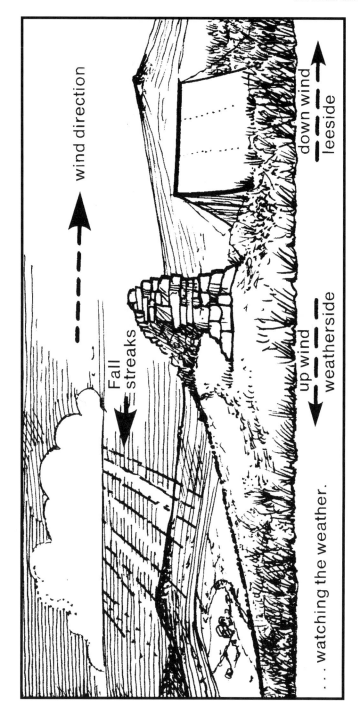

... watching the weather.

Permissible to camp on the site - all land belongs to someone. Permission to camp should always be sought from the landowner. In Scotland it is a criminal offence to camp without permission and this is true in certain countries on the Continent. There are, however, certain areas, usually in the wilder remote upland areas of England and Wales, where the tradition of free camping is tolerated by the landowners.

Free from objective dangers - it is vitally important that camp sites should be free from dangers which may take a variety of forms ranging from flash floods to cattle. Sharing a camp site with any form of livestock other than sheep is asking for trouble.

Able to provide some degree of privacy - mobile campers only have the seclusion of a small lightweight tent and, of necessity, their ablutions must take place in the open. Some degree of seclusion and privacy is essential both for the campers and for the inhabitants of any nearby dwellings.

Within easy reach of water - the site must be reasonably near water, but lightweight campers are usually prepared to carry their water a little way if it enables them to choose a better site which fulfils more of the criteria listed above.

Reasonably level - the ground on which the tent is pitched needs to be as level as possible within the limits of the terrain. Sleeping on a slope can be very uncomfortable as there is a tendency for all the occupants of the tent to finish in a heap at one side, or end, of the tent. If the tent has to be pitched on a slope, then it is better to sleep with the feet facing downhill.

Able to provide a secure anchorage for the tent - the tents which blow away are the ones which have not been fastened to the ground properly. The ground where a tent is pitched must provide a secure anchorage for the tent pegs. For many lightweight campers the site may be Hobson's choice! Sand, semi-marsh and stony ground provide the greatest problems and ingenuity may be necessary to overcome the problems.

Pitching the Tent

With such a variety of tents available, it is impossible to suggest a routine for pitching a tent, but one can say with certainty that a well-pitched tent is a safe and secure one. Objects which may puncture the ground sheet must be removed, and anything which can be done to improve the surface will increase the chances of a good night's sleep; you may well lie awake all night wishing you had devoted more attention to the ground you are sleeping on.

Guy lines should be run out in line with the seams and, if the tent has main guy lines, heavier pegs should be used. If the ground is loose or very soft, then stones should be placed on top of the pegs, care being taken to ensure that they will not fray the lines. This is where the extra trouble of adding a selection of different tent pegs of varying shape, length or material to those supplied with the tent gains its reward. It is often possible to fashion additional pegs or stakes out of twigs or wood. The alloy or steel pegs used with lightweight tents must be driven into the ground up to the hilt and at an angle of 45% to the surface of the ground with the head away from the tent. The tent should normally be pitched with the entrance facing away from the wind, but if it is pitched close to a windbreak, then it may need to be the other way round. Tents should not be pitched immediately under trees. Two people should never take more than 15 minutes to pitch or strike a lightweight tent.

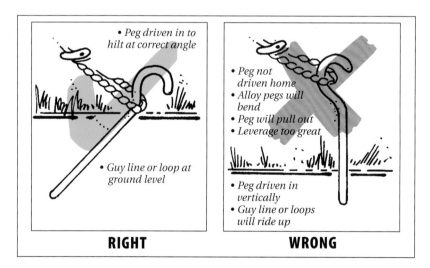

Living in a Tent

When two or three people are living within the confines of a few square metres of tent, organisation and tidiness are essential, especially in wet weather. It is important that the occupants establish a routine for 'who does what'. This should be sorted out before reaching the camp site in the evening. No matter how tired and weary the group may be, tents should always be pitched straightaway. Immediately after the kettle has been filled and placed on the stove to boil for the 'brew-up', the tent should be pitched and, with good teamwork, it should take no longer than the kettle takes to boil. A hot drink can do much to restore morale when a group is cold and wet, as well as remedying any dehydration. After a drink, preparations can be made for cooking the evening meal.

After the meal, equipment which is not in use should always be packed away. Stoves and utensils which will come to no harm if they get wet are best stored outside near to the tent. Experienced campers prepare for the unexpected. Pans and dishes should be washed after the meal, the site tidied up and all equipment restored to the pack, except for items which will be required overnight. The torches should always be to hand beside the sleeping bag. Boots, or outside footwear, should never be worn inside the tent but should be placed inside the tent on either side of the door. The lightweight groundsheets must always be treated with great care otherwise they will become porous; bare or stockinged feet must be the rule. Every effort must be made to keep the inside of the tent dry in wet weather by leaving wet over-clothing under the fly sheet, if there is one, and there should always be a cloth to hand to wipe up any spillages. One set of dry clothing must always be retained even if this means donning wet clothing in the morning.

When striking camp, the underside of the ground sheet must be wiped clean and dry and the tent pegs cleaned with some rag which is always stored in the peg bag. If the tent is wet, then it must be shaken to remove as much of the water as possible before packing. A wet tent must always be hung out to dry on return to home or base.

Hygiene

Hygiene is a most important aspect of camp craft and is of major concern to both Supervisors and Assessors when they visit a camp site. Personal cleanliness should always remain at a high level throughout the venture. Face, hands and feet should always be washed at the end of every day and teeth should be cleaned. Every effort should be made to wash socks as frequently as possible and they can be dried while on the move if the weather is fine by fastening them to the outside of the rucksack.

Water supplies must be kept clean. **No washing should take place in streams and dirty or greasy water must never be thrown back into a stream.** Dirty water should be poured into a hole in soft ground, well away from the steam, made by removing a piece of turf with the trowel.

All litter, tins, bottles, paper and remnants of food **must be removed from the camp site.** It may be possible to make arrangements with the landowner concerning its disposal. If not, it must be carried by the group until it can be deposited in a bin. Sometimes Assessors and Supervisors can be persuaded to take it away with them. Spare plastic bags should always be carried for this purpose. It is neither acceptable to burn rubbish, as it always leaves an unsightly patch, nor to bury rubbish as it will always attract animals.

When meeting the camp site owner on arrival at the site, discuss toilet arrangements. If none are available, dig a latrine well away from any stream and any place which might be used by other campers to pitch their tents. Using the trowel, remove the turf in one piece and dig a hole at least 20 cm deep (8 inches). Replace the turf after use so that there is no trace left. Do not remove rocks and then replace them as this leaves the site unusable for other campers.

Leaving the Camp Site

Before the camp site is vacated, the landowner should be thanked and any payment made. Any stones used under pans to prevent the grass being scorched, or used to secure guy lines, must be replaced. The site should be searched to ensure that it is clean and free from litter; even a match stick counts as litter.

After a camp site has been vacated there should be no trace of its ever having been used for camping; whether the site was on the immaculate lawn of a country house or on the top of some lonely fell. Good campers leave nothing behind but their thanks.

In conclusion, all campers should take a pride in being able to remain clean, warm and dry, well-fed and, above all, comfortable under all the conditions they are likely to encounter.

Joe Cornish

3 · 3

Catering
for
Expeditions

xpedition catering must not be confused with camp catering. Camp catering, especially where a motor vehicle is used, can include fresh vegetables, meat, dairy produce and the local delicacies or regional specialities which all add to the enjoyment of camping. Expedition catering for an unaccompanied, self-reliant venture which is dependent entirely on the participants' own physical effort and where all the provisions have to be carried, presents an entirely different form of challenge. Here the task is to cram the greatest amount of energy into the lowest weight and bulk.

The need for a balanced diet becomes increasingly important as the length of an Expedition increases. For journeys lasting up to three or four days, normal eating habits will ensure that the diet is adequately balanced. Usually too much thought is given to achieving variety in the menu instead of paying attention to the need for an adequate intake of food and liquids.

No tables of carbohydrates, proteins or fats are included as campers ignore them, but excellent books and tables do exist, and the labelling of foods has improved greatly in recent years.

For three or four day Expeditions, the prime need is to pack as much energy or calories into the least weight and volume. Calories, or to be more precise, kilo calories - kcal: 1 kcal = 4.18 kilo joules - kj. The unit used in food labelling is the larger kilo calorie.) This is usually achieved by increasing the amount of carbohydrates (sugars and starches) and fats, and by using dehydrated food. Such advice may appear contrary to modern dietary advice, but no harm will result over the duration of a venture as all the energy will be burnt up and participants should be fitter and healthier at the end of the journey than at the beginning. It is not so much a case of seeking additional fats but more of not avoiding them, by eating fish in oil, luncheon meats, preserved continental sausages such as salami, and butter or butter substitutes, on bread or biscuits. If too much food is carried, weight will be a handicap; too little food will result in hunger and physical efficiency may be impaired. Three, four, or five thousand calories or more may be needed each day on an Award Expedition.

As a rough guide:
- Simple carbohydrate - the sugars - provide energy very quickly.
- Complex carbohydrates in the form of rice, pasta, bread etc. supply energy over the medium term.
- Fats provide energy over a longer period of time.

Cakes, pastries and biscuits are a mixture of simple and complex carbohydrates and provide energy in both the short and medium term. Proteins do not play an important part in three or four day ventures as there is an ample supply in a normal Expedition diet and the body prefers to derive its energy from carbohydrates and, to a lesser extent, from fats.

Expedition catering is usually carried out in tent units, which usually means in pairs, though it may mean three or occasionally four. The menu, the food and the planning should be based on the tent unit.

Only take foods which:
- Contain the greatest amount of energy (kcal/kj) in the smallest weight and volume.
- You and all your cooking partners like and enjoy.
- Are simple to prepare 'one-pot meals', such as stews, curries and pastas.
- Cook quickly to save fuel.
- Will keep, especially in hot weather.

Supermarkets carry an endless variety of dehydrated foods and meals so there is no need to carry water around masquerading as food. The packaging of these products in foil or plastic is airtight and durable, usually waterproof, and stands up very well to buffeting of Expeditions. The outer wrappings or cartons should be removed to reduce bulk, rubbish and weight but make sure that the cooking instructions are retained.

The preparation of food is simplified and fuel can be saved if foods are carried which are liked by everybody in the tent group, and the menu prepared in advance with all participating in the process. Cooking time is reduced, there is no need to duplicate the cooking or to carry around a great assortment of food bags and packages.

The vast majority of dehydrated foods also have the advantage of being quick and easy to prepare, needing only the addition of water, and cook very quickly. The quantities will need to be increased; if the packet indicates that it will feed two, assume that it may feed one on an Expedition. Take foods which are simple to prepare, 'one-pot' meals such as stews, curries and pastas are ideal. You will not have the problem of trying to keep food warm if the menu is uncomplicated: cook it - eat it! Variety is of no great importance in a journey lasting three or four days. As Cervantes states in 'Don Quixote': *"Hunger is the best sauce."*

Check how long the food takes to cook. The total weight of the catering element in one's pack is **the weight of the food plus the weight of the fuel required to cook it**. Choose dehydrated foods which cook quickly. Some soups are 'instant', while other packet soups may require 5-15 minutes' boiling before they can be consumed. There is a whole range of pre-cooked noodles, potatoes, peas, rice and pastas which only take five minutes to prepare against the 15-20 minutes of the traditional product. Careful selection of dehydrated foods can greatly reduce the amount of fuel used on a venture.

Expedition foods must be chosen which will not go bad during the course of the venture, especially in hot weather. This rules out fresh meat. Sealed pre-packed bacon may last until the morning of the second day. Dried, cured or smoked continental meats and sausages make ideal Expedition food and they will normally last for the duration of a venture. There is also a wide range of vegetarian foods available, usually based on soya, which have excellent keeping properties.

Tinned foods do not feature to any great extent in three or four day Expeditions, but there are a few which are worthy of very serious consideration. Corned beef, luncheon meat, sardines, pilchards and tuna all contain a great deal of fat or oil which gives them a very high energy content and makes them a valuable addition to any Expedition menu. The weight of the tin in relation to the energy supplied is negligible; they are waterproof and the contents will last for years.

General Considerations

Plan meals ahead, experiment with dehydrated food meals at home and on practice ventures. Foods can be weighed and measured out in advance. Experiment with the various powdered milks to see which ones can be boiled. Plastic bags provide suitable storage for most foods, but butter, jams and spreads are messy unless kept in plastic containers or refillable plastic tubes which are now available on the market. Plastic/metal ties frequently puncture plastic bags and rubber bands are a safer way of securing all bags. All food must be packed and waterproofed so that it will stand up to the hammering it will receive in rucksacks or containers which are dropped, sat on and squashed during the course of an Expedition. Drinking mugs may be converted into handy measures by scratching a few lines in the right place.

Drink

Many participants place far too much emphasis on planning the menu and on food, and far too little on the fluid intake. It is possible to survive for a considerable number of days without food, but ill-effects occur and efficiency is impaired in a comparatively short time if fluid intake is inadequate. A large fluid intake is essential in all ventures where there is a high expenditure of physical energy, particularly when walking, cycling, canoeing or rowing. Much of the fatigue which participants experience towards the end of the day is frequently due to dehydration. When you have a very bad thirst it is probable that your physical efficiency is already impaired. Drinking becomes even more important in hot weather, whether in Great Britain or abroad, when it may help to delay the onset of exercise-induced heat exhaustion.

It is difficult to quantify how much liquid is needed during the travelling part of the day as so much depends on the weather conditions, the individual and the amount energy expended. Several litres may need to be consumed during the journeying and mid-day break. In very hot, dry conditions, people who are not acclimatised to the heat or used to

Paul Smith (Halina/Fuji Bursary)

More attention should be paid to liquid intake

coping with a very high workload may well need a total fluid intake of 7-8 litres (15-16 pints) every 24 hours.

Much of the body fluid is lost through sweating when performing a physically demanding activity, especially in hot weather. It is necessary to replace not only the fluid loss, but the salt which is also excreted. There is no need to carry salt tablets in Britain or the rest of Europe, but food should be well salted and plenty of salt should always be carried. Sadly it is all too frequently forgotten when it is most needed, while at home there is frequently an excess of salt in our diet.

There is a tradition amongst many experienced climbers, hill and mountain walkers not to drink on the hill, or they drink from mountain streams. It is a policy which can no longer be recommended to young people within the Award Scheme. It may be safe to drink from the clear mountain streams of the Scottish Highlands, but over practically all of England, most of Wales and Northern Ireland the situation is very different. Even if a dead sheep is not lying a hundred metres upstream from where you take your water, there is a probability that the water may be contaminated. Expeditioners are advised to go through, rather than over, wild country and there is more chance of the water being contaminated in the lower areas where there is a greater intensity of recreational use and increased agricultural activity. For ventures in normal rural and open country, where the vast majority of ventures take place, the problems are even greater with the contamination by nitrates and slurry from farming, as well as sewage and industrial effluents.

Water intake for ventures on foot may need to be in the order of five, or even eight, litres a day and this intake cannot just take place only in the morning and evening. Ventures in the hot weather of continental Europe may pose even greater problems and the needs of canoeists and sailors on rivers or salt water must not be overlooked.

Participants should 'tank-up' with fluid before departure from the camp site in the morning and **drink as much as possible as soon as possible** on arrival at the camp site in the evening. A suitably sized plastic container with a screw top will serve as a water bottle and help to overcome the problems of dehydration during the day. Water should be sipped in small amounts at regular intervals during the journey to derive the most benefit from a limited supply. No opportunity to slake ones thirst should be missed during the journey by drinking copiously. A lunch time break provides an ideal opportunity to prepare a 'brew', even in the hottest weather. If the break is taken close to a stream, water can be sterilised by boiling and then drunk as tea or coffee, while extra water can be boiled, allowed to cool and then used to top up the water bottle.

Water sterilisation tablets are available from chemists. The directions accompanying the tablets should be followed and any taste disguised by adding a small amount of lemonade or fruit drink powders. To ensure that there is sufficient to drink, especially in hot weather, always carry more tea, coffee, sugar and milk powder than you think you will need. Fruit flavoured breakfast drinks and lemonade powders are available and a search of the supermarket shelves is worth the effort.

COOKING

There is progression in the catering syllabus of the Award Scheme from being able to prepare a simple meal at Bronze level, to the ability to prepare and use dehydrated foods at Gold level and to make substantial meals under camp conditions.

An introduction to the variety of stoves on the market is made in Chapter 3.4 - Cooking Stoves.

Breakfast

Individual preferences are always a most important consideration in camp diet. Many campers begin the day with a substantial breakfast and, for some, no day is complete without bacon and eggs. This has much to commend it as the fat and the protein stay in the stomach and not only provide energy over a long period of time, but stave off the pangs of hunger. Others prefer to avoid washing greasy pans and stick to cereals, muesli or porridge variants followed by biscuits or bread and jam. Whatever the preference, it is good practice to stoke up with plenty of food and drink before setting out in the morning.

Lunch

Attitudes to lunch vary. Some groups are content to have pockets full of nuts, dried fruit and biscuits which they supplement with chocolate bars and boiled sweets. These are concentrated forms of energy and may be consumed while on the move, a form of 'drip feed'! It is possible to meet all the energy requirements by following this practice, which keeps blood sugar levels high over a prolonged period. It is cheaper to buy a variety of dried fruit and nuts by the kilo/pound and mix them yourself. Others prefer their lunch break with sandwiches and a 'brew'. Many like cheese and biscuits or biscuits with some spread such as jam or peanut butter, while others would not be without their loaf of bread. It is necessary to 'stoke-up' with fuel during the journey and, whatever method is adopted, an energy boost is provided which is essential for the body to work efficiently where long-term stamina is important.

Evening Meal

Award conditions require that one substantial meal be cooked each day and this is nearly always the evening meal. Even when limited by one stove and a couple of pans it is possible, with a little practice, to prepare hot, filling three-course meals in a very short time. Such a meal might consist of soup, a curry, stew or pasta followed by a hot or cold pudding or cheese and biscuits with coffee. In addition to the dehydrated foods mentioned earlier, there are whips and mousses which only require the addition of water or milk, and ground rice and semolina preparations which need boiling milk to turn them into hot nourishing puddings.

Tony French (Halina/Fuji Bursary)

'One-pot meals' are ideal

When the tent has been pitched, the water fetched and you are ready to prepare your evening meal, lay out all the ingredients and pans in the order in which they will be needed. Then, and only then, light the stove. Lids on pans help to conserve heat; food will cook more quickly and fuel will be saved. Unless the cooker is a modern spirit stove which is well shielded, a wind shield will improve efficiency and save fuel. The milk for the pudding might be boiled first, the dish prepared and set aside for later; then the soup should be made and drunk from mugs while the main course is cooking. Immediately after the main course is prepared, water can be boiled for the coffee and washing up. If drinking water cannot be obtained from a farm then it may be necessary to boil water, set it aside to cool and then fill the water bottles for the following day. After the meal, the experienced camper will wash up, tidy up and then pack away any equipment not needed. A hot drink and biscuits are usually sufficient before going to bed, but do not forget to eliminate any dehydration which may have occurred during the day!

Cooking Stoves

A ccidents in the Expeditions Section are fortunately very rare, but the few which do occur are mostly associated with stoves and cooking. In every reported instance, the accident has occurred through a failure to appreciate the dangers of handling highly flammable liquids and gases.

Open fires (camp fires) are not an option for Expedition camping, either for cooking or social enjoyment, because of the environmental damage they cause. Landowners will not permit them, not only because of the unsightly scars that they leave on the ground, but also because of their previous experience with damage to trees, hedges and fences in the search for fuel. You are more likely to be successful when asking for permission to camp if you assure landowners that you will be using stoves and there will be no open fires.

There is a wide variety of stoves available for participants on their ventures, though the choice of fuel is usually limited to either methylated spirits or gas cartridges.

Petrol stoves are not popular with Operating Authorities because of the dangers associated with the fuel and the regulations concerning its storage, though in skilled and careful hands they are very reliable and efficient. Spirit stoves account for around 80% of the market, but gas stoves of an improved design are making a come-back. Paraffin pressure stoves, such as the Primus or Optimus, are rarely seen, though many devotees still regard them as the safest, most reliable, efficient and economical stove available, providing one knows how to operate them in all conditions.

Gas stoves, with a marked improvement in their design and self-sealing cartridges, are increasing in popularity. Self-sealing cartridges have brought about a significant increase in the safety of gas stoves.

The greatest care and discipline must be exercised at all times when using stoves as the accidents which do occur are invariably the result of carelessness and the failure to follow a few simple rules which must always be obeyed.

Spirit Stoves

The popularity of the spirit stove, which serves as a stove and canteen, is easy to understand. They are well-designed, light and compact, with the stove and the canteen all packing together in a single unit. They are stable with their broad base and moderate height and do not scorch the grass. Simple to use and clean, free from the oil and odours associated with petrol and paraffin stoves, they are easy to light and use fuel efficiently - 50g of fuel will boil around a litre of water. Stoves are easy to shield from the wind and it is possible to regulate them to a certain extent by turning them around in relation to the wind direction and using a built-in regulator ring. They are relatively inexpensive, fuel is readily available and not too expensive when purchased in bulk, and

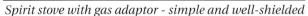

Spirit stove with gas adaptor - simple and well-shielded

THE SAFE WAY TO FUEL AND REFUEL SPIRIT STOVES

• *Small fuel bottle*

• *No large capacity containers*

1 GALLON PLASTIC CONTAINER

• *Away from tents and naked flames*

• *Not in strong sunlight*
• *Allow to cool*
• *If too hot to handle it is too hot to fill*

they do not have the restrictions associated with the storage of petrol. When used with care, spirit stoves are capable of providing years of safe and trouble-free service.

Methylated spirits is available on the Continent under a variety of names: France – alcool à brûler; Germany – denaturierte spiritus; Holland – brandspiritus; Spain – alcohol desnaturalizado or metanol and Italy – alcool denaturato.

Spirit stoves are so simple to operate that training is frequently not as thorough as it used to be with paraffin and petrol stoves. The few accidents which do occur are usually associated with spirit stoves though there are many hundreds of thousands in use. The dangers arise from the fuel. **Methylated spirits is highly volatile, has a low flash point and in strong sunlight burns with a virtually invisible flame.** Many people fail to appreciate how quickly the volatile vapours can spread. There have been instances when fuel has been added to a stove which was still too hot or the spirit has been added to a stove in strong sunlight where the practically invisible flame has not been detected. Such action may lead to disastrous consequences with the fuel container catching

Stoves must be well away from tents

D.W. Elson

fire and acting as a flame thrower. Tents have been burnt down with injuries to the occupants, or companions in the vicinity have been burnt. Practically every accident which has occurred has been the result of a failure on somebody's part, with the person who has caused the accident escaping without injuries. **These accidents must be eliminated by better training and a disciplined, responsible procedure in the use of stoves and their fuel.**

A safety routine for spirit stoves:

- **Stoves must not be filled from a bulk container such as a one gallon can or plastic container. (It is permissible to sell and store methylated spirits in a plastic container).**
- **The spirit must be carried in one or more bottles specifically designed for the purpose. These should normally be around half a litre in capacity and the stoves should be filled from these bottles.**
- **One person should carry the stove or the spirit cup downwind, away from the tents and it should be fuelled, or refuelled, in a place where there are no naked flames in the vicinity. The screw cap on the spirit bottle must be replaced immediately and the bottle returned to a safe place. Some bottles specially designed for carrying fuel have a pouring hole in the screw cap which only needs to be slackened to fuel a stove. Pouring spouts are available for some makes of fuel bottles.**
- **Do not refill a stove until you are sure that the flame is completely extinguished and the stove has cooled. If the stove is too hot to handle, then it is too hot to be refuelled. Remove the pan and carefully place a hand over the spirit cup or hold a sheet of paper just above the burner. This will indicate how hot the stove is as well as shading it from any strong light.**
- **The stove must be placed on a firm level surface, at least one metre away from the tent, in a place where it cannot be knocked over.**

Practical lessons should take place out of doors where possible and the above routine carefully observed. Where lessons have to take place indoors, only one stove must be in use.

Spirit stoves are directional with ventilation holes on one side. By turning the holes towards the wind draught is increased. If the cooker is exposed to a very strong draught, the flame may be increased to such an extent that it will melt the base of the stove. Conversion kits are available to convert some spirit stoves to gas. Those wishing to convert a spirit stove to gas should make sure that it will accept the conversion kit. Older spirit stoves may not.

Gas Stoves

In the past gas stoves have caused as many accidents as spirit stoves. This problem has been largely eliminated by the introduction of the valved, self-sealing cartridge. **All gas stoves used on Award ventures should be of the type which use self-sealing cartridges/cylinders.** Traditional gas stoves also suffer from design problems and there are many of them still in use. They are tall with the burner sitting on top of the gas cartridge and the pan on top of the lot; with their narrow base they tend to be unstable and easily overturned. They are very difficult to screen from the wind and a great deal of the heat is wasted. Modern gas stoves have been improved in design - they are lower, more stable and easier to shield from the wind. Modern windshields are made of heavy aluminium foil which can be bent to the required shape.

Gas stoves need protection from wind and draughts -
windshields are important

Gas stoves, with their disposable cartridges, are clean, reliable and extremely simple to operate. Butane cartridges do not vaporise very well in cold weather, but propane cartridges or butane/propane mixes are available. This should not be a problem during the Expedition season.

Self-sealing cartridges may be removed from the stove for travelling, and partly-used cartridges may be replaced by full cartridges before embarking on a venture. Cartridges which have been partly used should be marked with a spirit marker to avoid confusion.

Non Self-sealing Cartridges or Cylinders (valveless)

The use of valveless cartridges or cyclinders which are not self-sealing should be avoided in Award ventures. Many distributors no longer market stoves which do not use self-sealing cartridges.

An assortment of gas cylinders or cartridges with self-sealing valves

The danger with the older type of gas stove arose from the cartridges not being self-sealing. A spike in the burner head pierces the cartridge. Because so many of the older type of gas stoves and lanterns are still around, instructions for changing the non self-sealing cartridges are included in this *Guide*.

Instructions for changing pierceable cartridges without a self-sealing valve:

- Make sure the cartridge is completely empty by burning off any remaining gas.
- Close the valve on the burner by turning clockwise.
- One person should then take the stove away from tents, other people and any naked flames.
- Remove the burner head completely by unscrewing anti-clockwise and place on the ground.
- Remove the clips or the baseplate by unscrewing anti-clockwise, and then remove the empty cartridge.
- Insert the new cartridge, fasten the clips securely or replace the baseplate, twisting clockwise, making sure that it is properly and securely seated.
- Replace burner head by screwing on clockwise, making sure that it is not cross-threaded at the start.

Some fuel bottles are fitted with pourers which may be purchased separately

CHANGING PIERCIBLE CARTRIDGES
WITHOUT A SELF-SEALING VALVE

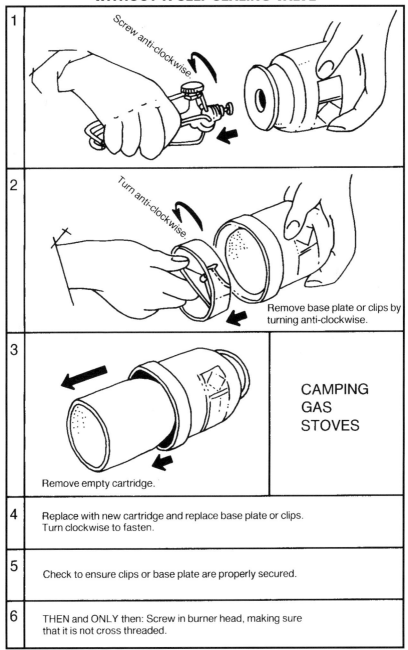

1 Screw anti-clockwise.

2 Turn anti-clockwise. Remove base plate or clips by turning anti-clockwise.

3 Remove empty cartridge.

CAMPING
GAS
STOVES

4 Replace with new cartridge and replace base plate or clips. Turn clockwise to fasten.

5 Check to ensure clips or base plate are properly secured.

6 THEN and ONLY then: Screw in burner head, making sure that it is not cross threaded.

Fuel Containers

Spirit, petrol and paraffin stoves will need fuel bottles. These should always be of metal or specifically designed for the purpose. An adequate supply of suitable gas cartridges will be needed for gas stoves. The weight of these can be spread amongst the group to reduce the load. While every effort must be made to keep the amount of fuel carried to the minimum and economy practised at all times, you will need to have sufficient fuel not only to cook your food, wash up and brew a generous supply of drinks, but possibly to sterilise water as well. It is essential that you determine, during practice journeys, how much fuel you will need to carry and keep a record.

A towel soaked in water will make a very effective fire blanket. Methylated spirits and water may be mixed so there is no danger of a fire being spread; the water simply cuts off the air supply and cools and dilutes the spirit. A wet towel is also effective with a fire involving gas stoves. A pan of water with a towel beside it is an excellent stand-by for dowsing a cooker fire in an emergency, even though it is pointless wetting the towel beforehand.

By always conscientiously following these drills, stoves can be used safely and confidently for camp cooking, providing there is never any complacency or carelessness.

It is an offence to carry fuel or cartridges on an aircraft.

Navigation: Map- Reading

The importance of good navigation for all Expeditions, especially those on foot, cannot be over-stressed as the majority of difficulties arise from groups being lost, hours overdue or reported missing are caused by inadequate navigation, together with a lack of physical fitness and overweight packs. Even if the Expedition does not have to be abandoned, it inevitably results in extra fatigue from increased distances having to be covered, wasted hours and possibly benightment, as well as increased anxiety for all those who are responsible for supervision. Competent navigators experience confidence, satisfaction, pleasure and a freedom of movement which usually extend into all aspects of their life in the outdoors.

The techniques outlined in the following four chapters apply equally to ventures on rivers and other inland waters and variations in technique may only be needed in estuaries and coastal waters.

It is impossible to provide the navigational detail and describe all the technical skills which would fill a book within the limitations of four or five chapters, and you are advised to refer to *Land Navigation, Route Finding with Map and Compass* published by The Duke of Edinburgh's Award.

THE PREPARATORY SKILLS

The skills in this chapter will not enable you to find your way around the countryside. Using a map to find your way depends on being able to relate the map to the countryside, find where you are on the map and 'set', orient or orientate the map. These skills can only be mastered by considerable practice out-of-doors. These preparatory map skills are essential as route finding is always based on good map-reading. They are best acquired indoors in bad weather to make more time available for work outside when the weather is better.

A map is simply an aerial picture, or representation, of the ground and shows all the important landmarks and features such as hills, valleys, rivers, roads, footpaths and buildings by means of signs and symbols. All the features are in the correct direction from each other and they are at the correct 'scale distance' apart. Maps are reduced representations or pictures of the ground they represent, and the size of the reduction is the scale of the map. In this book the words - 'ground', 'country', 'landscape' and 'terrain' will be used in the text and diagrams to denote the actual earth's surface on which we move and to differentiate this from the map.

MAP DIRECTION

Navigation is about direction and maps give the direction from one place to another. Direction is best expressed in terms of the more important points of the compass, the 4 cardinal (North, East, South and West) and 4 half-cardinal (North East, South East, South West and North West) points (see below). These eight directions are sufficient for all your needs. Relate the direction of any two places to the nearest of these points. Where greater accuracy is required it is more convenient to express direction in terms of degrees.

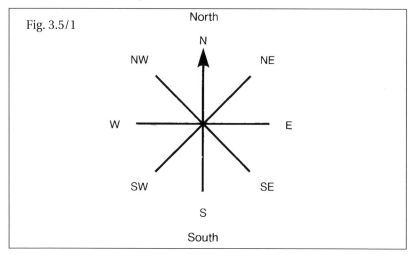

Fig. 3.5/1

The map itself provides direction; the top of the map is always the North, the bottom South, the right hand edge is the East and the left hand side indicates the West. Similarly the corners, or the diagonals, give the half-cardinal directions - North East, South East, South West and North West.

Fig. 3.5/2

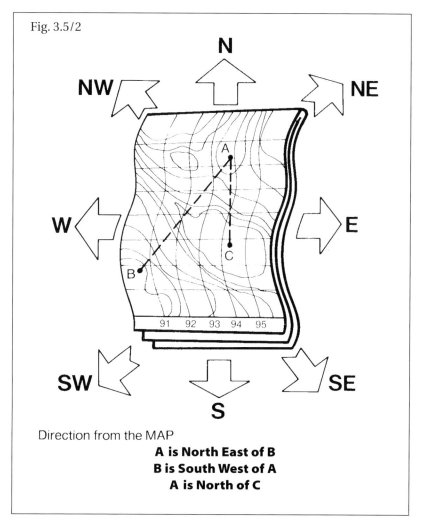

Direction from the MAP

A is North East of B
B is South West of A
A is North of C

Whenever two places are considered, the first and most important relationship which should be established is that of direction. Great care should always be placed on the use of the words 'from' and 'to', especially when communicating with others, or you may well find yourself travelling in exactly the opposite direction.

SCALE AND DISTANCE

Choosing a map of the right scale is vital for both effective navigation and in assisting the learning process. The larger the scale of the map the more room there is to show detail and the easier it is to navigate with, but a price has to be paid for this convenience. The larger the scale, the more maps are needed to cover any given area of country. This obviously adds considerably to the cost of the maps and you may need to carry several maps on your journey if you are travelling a considerable distance. If the scale of the map is too large you are continually 'walking off the map'. As a general rule, the faster you travel, the smaller the scale of map required. Walkers tend to use 1:25 000 and 1:50 000 scale maps, while a cyclist might find it more convenient to use a 1:250 000. A motorist will often use a map covering half the country.

Scale is expressed in three ways :

- **In words**. 'Four centimetres to the kilometre'. This is a very simple and easily understood way of expressing scale – a line four centimetres long on the map represents one kilometre of actual ground.

- **By a representative fraction or ratio**. A scale of 1:50 000 means that one unit of length on the map represents 50,000 of the same units on the ground. One centimetre on the map represents 50,000 centimetres on the ground, or more practically, two centimetres on the map represents one kilometre on the ground. A scale of 1:25 000, or 4cms to the kilometre, is referred to as 'one to twenty-five thousand'. Always remember that the larger the ratio, the smaller the scale of the map. The 1:50 000 map, where 2cms represents 1km of ground is a smaller scale map than the 1:25 000 where 4cms on the map represents 1km on the ground.

- **By a scale line**. This is a visual, practical and simple way of showing scale. It does not require any arithmetical ability or the manipulation of numbers. Scale lines are drawn on the margins of the map in the most frequently used units, kilometres and fractions of a kilometre or miles and fractions of a mile.

Choice of Scale

Two maps are of overwhelming importance to those involved in the outdoors in Great Britain, Ordnance Survey's 1:50 000 *Landranger* series and 1:25 000 *Pathfinder, Explorer and Outdoor Leisure* series. All who wish to become proficient in route finding in England, Scotland and Wales must be capable of using both of these scales with equal ease. The larger scale 1:25 000 map, with its greater detail, is the better map with which to learn to navigate, and every effort should be made to obtain one of your own locality. The whole of Ulster is covered at the 1:50 000 scale, as is most of continental Europe, but there are maps at 1:25 000 scale for many areas of the Continent. If it is impossible to obtain a 1:25 000 scale map, use a 1:50 000 as all the skills and techniques are identical.

Fig. 3.5/3 Comparison of Scales

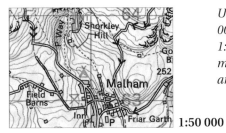

1:50 000

Use both 1:25 000 and 1:50 000 maps for route finding. 1:25 000 for learning to map read and planning and 1:50 000 for tracings.

1:25 000

Looking at the map extracts you see that, by doubling the scale of the map, there is four times as much area in which to show detail, thereby providing a much clearer representation of the ground, which is the first great advantage of the larger scale. The larger scale also enables field boundaries to be shown. The 1:50 000 map gives the impression of openness, tempting one to consider a route leading off to either side; comparison with the 1:25 000 immediately shows that the area is covered with small fields or meadows and deviation from the track is out of the question. This is the other great advantage of the 1:25 000 scale map; it enables more realistic routes to be planned, reduces the problems over access to land and makes for happier relations with the people who live and work in the countryside.

Distance

If direction is the first and most important relationship which the navigator has to establish between any two places, the second is distance. It is vital to be able to measure the distance between two places on the map quickly and accurately. Rarely is this in a straight line or 'as the crow flies'; it nearly always involves measuring around the twists and turns of tracks and footpaths. The most usual method is to take a piece of thin string and lay it carefully along the road or paths which form the chosen route and then lay it straight along the scale line on the margin. There are usually two scale lines, a metric one in kilometres and metres, and one in imperial measure. Note carefully where the zero is on the scale; to the right of the zero are whole units while to the left are the smaller divisions or fractions. By positioning one end of the string on the scale line at the appropriate whole unit to the right of zero it is possible to read off the distance accurately from the smaller fractions to the left of the zero. See Figs 3.5/4 and 3.5/5.

Fig. 3.5/4

Lay String or Edge of Paper along Route

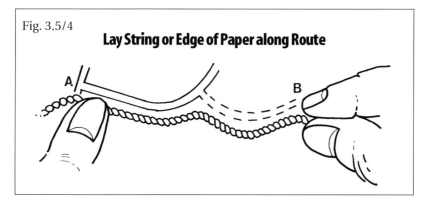

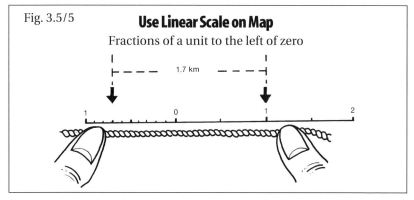

Fig. 3.5/5 **Use Linear Scale on Map**
Fractions of a unit to the left of zero

1.7 km

Instead of using a piece of thin string, it is possible to use the same technique with the edge of a piece of paper, marking the paper at each sharp turn and then laying it along the scale line.

A Measuring Scale

As an alternative to following the route with the edge of a piece of paper, it is possible to construct a measuring scale out of a piece of card and read off the distance directly instead of having to make use of the linear scale on the margin of the map. With a little practice it is possible to become very adept at measuring distances in a matter of seconds with far less chance of error than using the scale on the map or the measure on your compass.

As well as the ability to measure distances quickly and accurately on the map, it is helpful to estimate distances by the eye alone. Practice measuring distances with string, paper and your measure until you are proficient, even over intricate routes. Before you measure the distance between two places, estimate the distance first by the eye in a direct line and then along the chosen route.

Commercial Map Measurers

It is possible to purchase an instrument called an opisometer for measuring distances on a map. Choose one with a straight shaft as they are easier to twirl between the finger and thumb when following the route. Check that the needle on the dial is at zero, then trundle it along the route you wish to measure. At some scales it is possible to read the distance directly off the dial, but usually it is necessary to trundle them in the opposite direction along the scale on the edge of the map. With experience in manipulation, an opisometer can be a great help in speeding up route planning.

Distance and Time

Measuring distance on the map is only a means to an end. The real purpose is to determine how long it will take to get from one place to another. By adding the time the journey should take to the time of departure it is possible to estimate the time of arrival (ETA) at our destination. Considerable practical experience is needed before reliable estimations of speed of travel over various types of terrain can be made.

The 1:25 000 being twice the scale provides four times the area to show more detail and, in particular, the field boundaries (see extract on page 128). The extra detail makes route finding easier, enables the correct path to be found and followed, and reduces the possibility of friction with the landowner.

CONVENTIONAL SIGNS

Conventional Signs are the shorthand of the map maker. These symbols are used to convey information without obscuring or cluttering up the map. They become familiar with constant use, but every time you meet one which you do not know look it up in the legend or key on the margins of the map. Compare the symbols on the 1:50 000 with those on the 1:25 000 as you should be able to change from one to the other without confusion. Conventional Signs are grouped into logical categories on the map margin but, for navigational needs, it is more helpful to divide them into three groups based on their relevance to route finding needs.

Point or Spot Features

This group contains the signs and symbols which represent particular places enabling you to pinpoint a location and establish where you are. Churches with spires or towers, all buildings, milestones or posts, television or radio masts, bridges, towers and tall chimneys are all obvious landmarks.

Line or Linear Features

This category includes all the features which have length: roads, paths, streams, rivers, a range or line of hills, electricity transmission lines and the less obvious features such as field boundaries. Because they have length they have direction, even a river meandering over its flood plain has a general direction and this can, in turn, provide you with direction.

Area Features

In the third group are placed all the features which have area. These are principally concerned with the type of terrain and vegetation. Obvious

Tony French (Halina/Fuji Bursary)

examples are woods, forests, scrub, sand dunes, mud flats, marshes, scree and boulder strewn land. These features may help in determining where you are but, more importantly, they all have a considerable influence on your speed of travel. Try and place all Conventional Signs into one of these three categories.

MARGINAL INFORMATION

In addition to the key to the Conventional Signs and the linear scales mentioned earlier, there is other very important information on the borders of a map. There will be a date, or dates, stating when the map was compiled and/or revised, and possibly the dates when the surveys from which the compilation was made were carried out. Some features will be out of date before the map is even printed. This is especially so around towns and cities. New features appear while others disappear; the camp site you were hoping to use may have been replaced by a four star hotel or have sunk without trace beneath a reservoir. Pay particular attention to the date on your map for it may help to account for many surprises and avoid much confusion. Information on the adjoining sheets and the National Grid is helpful.

GRID REFERENCES

Grid references do not help you to find your way around the countryside but they are an important means of communication. It is essential to be able to refer to a place on a map precisely and communicate its position quickly and accurately. The National Grid Reference System enables us to do this. A six figure grid reference will meet all your navigational needs. Maps are covered with blue grid lines running North/South called 'Eastings' and lines running East/West called 'Northings'. These are spaced at 4cms and 2cms apart on the 1:25 000 and 1:50 000 scale maps respectively, so that the distance between them always represents 1 kilometre of actual country. Each line is identified by two figures ranging from 00 to 99. In a six figure grid reference the first three figures indicate how far to the East the place is, while the second three figures show how far North the position is. The first two figures in each group of three are read from the map, while the third figure is estimated or measured. If you remember that the letter 'E' comes before the letter 'N' in the alphabet, you will have no difficulty in remembering that 'Eastings' come before 'Northings'. East then North! (See below).

As the grid lines are numbered 00 to 99 and each line is 1km apart, a six figure grid reference used without further qualification will be repeated

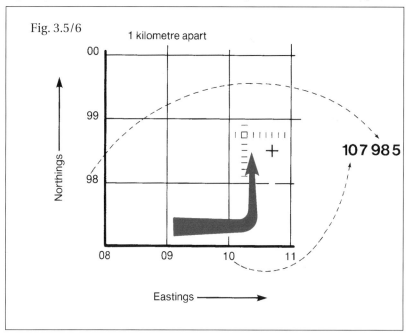

Fig. 3.5/6

every 100kms (about 62 miles), but this does not cause any confusion in practice. A six figure grid reference locates a place to within 100 metres. A scale called a 'Romer' can be used for measuring grid references accurately. These can be purchased, but it is much cheaper and far more instructive to make your own; one for the 1:25 000 and one for the 1:50 000 map on the same piece of card. Always estimate the grid reference by eye before measuring. If you are ever given a grid reference which does not make sense (such as the camp site appearing to be in the middle of the North Sea) try changing the last three figures with the first three.

RELIEF

The height, shape, land forms and steepness of the ground are shown on the map by means of contours. To be able to visualise the shape and nature of the landscape from the contours of a map may take many years of constant practice, but you will be able to acquire the three most important skills with little difficulty. These skills are:

- Finding the height of a place.
- Appreciating the steepness of a slope.
- Recognising a few of the major land forms such as a hill, valley, spur, ridge and high and low flat ground.

Contours

A contour is a line drawn on a map joining all the places the same height above sea level. The difference in height between two successive contours is known as the vertical, or contour, interval. On 1:25 000 maps the contours are at intervals of 5 metres in lowland areas and at 10 metre intervals in the upland areas. The contours are all at 10 metre intervals on the 1:50 000 *Landranger* maps. Every fifth contour is thickened to make it easier to follow, count and read. The numbers always read 'uphill'. (See Figs. 3.5/6, 3.5/7, 3.5/8 and 3.5/9).

If you are using a different scale map, or maps of a different country, the contour interval will almost certainly be found in the marginal information. All the procedures are the same.

Height

The ability to find the height of a place or, more usually, to be able to find the difference in height between two locations, is the third important relationship between two places after direction and distance. The

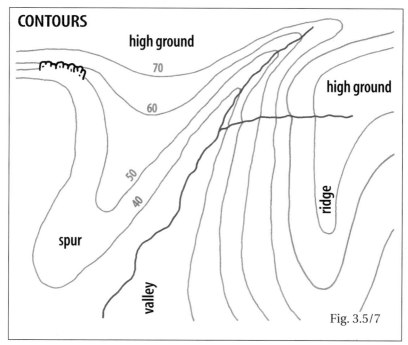

CONTOURS

high ground

70

high ground

60

50

40

spur

ridge

valley

Fig. 3.5/7

spacing of the contours indicates whether the slope is concave or convex which affects visibility.

Steepness of Slope or Gradient

The closer the contours are together, the steeper the ground. Flat land, whether it is high or low, will have few, if any contours. You can form an impression of how steep a slope is by how tightly the contours are packed together and from the 'orangey' colour of the map. But this impression

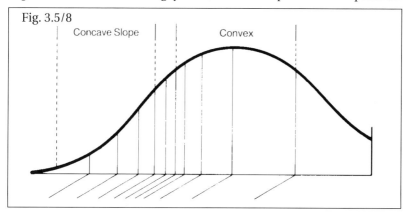

Fig. 3.5/8

Concave Slope

Convex

from the map alone will have little meaning for the novice until it has been translated by a great deal of practical experience. Later, by counting the number of contours per centimetre of map and comparing it with the country over which you are walking, you will soon be able to appreciate the steepness of slopes from the map. Fig.3.5/9 represents one of the simplest land forms, a hill or a knoll. The contours are closer together on the East than on the West, indicating that the slope is steeper on the Eastern side.

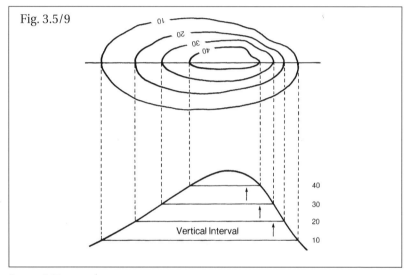

Fig. 3.5/9

Land Forms

In order to plan sensible routes you will need to be able to recognise certain common land forms from the map such as hills, valleys, spurs and ridges; there are others in mountainous country, but these will do to start with. Widely spaced contours, or where they are practically absent, indicate fairly level ground, a plain if it is at a low altitude, a plateau if the ground is high. All these land forms have characteristic shapes when represented by contours on a map. By constantly comparing map and landscape you will soon become familiar with them and learn to recognise them instantly. Fig. 3.5/9 shows the basic representation of a hill. In Figs. 3.5/10 and 3.5/11 the contours depicting a spur (high ground projecting into lower ground) have a resemblance to those of a valley (low ground between higher ground) so it is important to look at the contour heights. Usually, but not always, valleys have the blue line of a stream or river running through them.

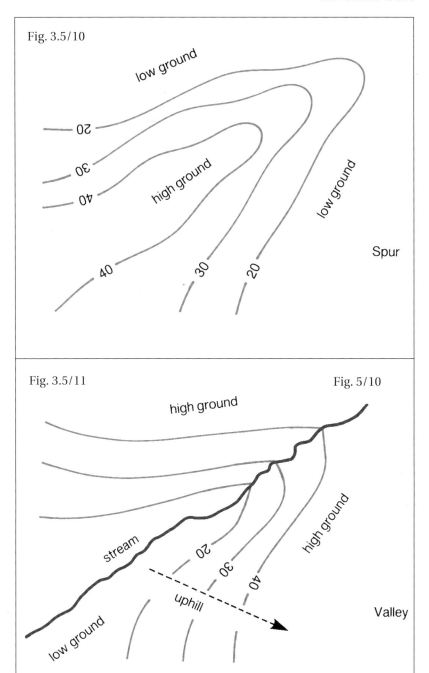

Fig. 3.5/10

low ground

20

30

40

high ground

low ground

40

30

20

Spur

Fig. 3.5/11

Fig. 5/10

high ground

stream

20

30

40

high ground

uphill

low ground

Valley

MAP READING

So far we have considered six aspects of map reading: direction, scale, distance, Conventional Signs, marginal information and contours as isolated skills. These skills must be brought together and blended into the art of map reading.

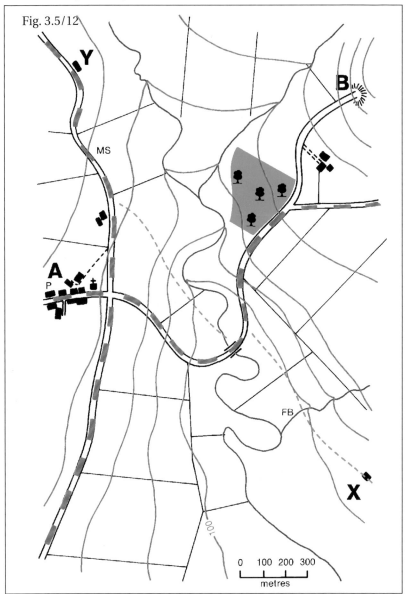

Fig. 3.5/12

Look at Fig. 3.5/12 and assume that you wish to travel from the Post Office at 'A' to a camp site at the quarry 'B'. From the map it is possible to give the following description of the journey:

"General direction NE . Distance 2.4 kilometres. I shall turn left outside the Post Office, towards the East and along the village main street. After 200 metres there will be a church on my left and after another 100 metres a crossroads. I shall go straight on at the crossroads and the road will descend a hill and veer to the right. After a sharp left-hand bend I will cross a bridge over a stream or river. The road will turn to the North and ascend the other side of the valley. On my left, the North West side of the road, there will be a wood. 200 metres beyond the start of the wood the secondary road will turn to the East but I will continue straight ahead towards the North East along an unmetalled road or track. After 300 metres, an unfenced track on my right will lead to some buildings, probably a farm. The unmetalled track or road will continue with a gradually increasing gradient for a further 300 metres until it reaches a quarry in a hillside and my camp site."

Note that it is personalised. Once you are able to map-read it will be possible to describe a journey between two places on the map as if you were actually travelling along the route. You will be able to think of features as being 'on my left', or 'straight on'. The first concern is with direction in terms of the cardinal and half-cardinal points. Because routes on land twist and turn, the general direction is sufficient for our needs.

After direction comes distance: the total distance along the route has been measured as well as the distance between important landmarks. These distances are a means to an end, for later on you will be able, by practical experience, to translate these distances into travelling times. The 2.4kms may be turned into a 25 minute walk.

Notice how emphasis is placed on the landmarks on the route, primarily to ensure that we are on the correct route and secondly to keep track of our progress along the route.

More detail could have been given in the description. From the Conventional Signs we know that the church has a tower and the wood is deciduous. If the route was longer it might be necessary to omit some of the detail. The amount of detail given in such a verbal description is not important. It is important that we are able to extract from the map all the detail that the map-maker has provided to enable us to find our way.

Choose two places a few kilometres apart on the map and write down descriptions of the most suitable route, using your own words, or alternatively describe the route to a companion. Practise this skill until it becomes second nature because it is what map-reading is all about. Once you have mastered this skill you will be able to regard yourself as a competent map-reader. With this ability you will be able to take a map of unknown terrain at the opposite end of the country, or the opposite side of the world, plan a route with all the essential navigational detail and landmarks. You will be able to visualise the sequence of their appearance. As your practical experience broadens to include different types of terrain and country, so you will be able to fill in the background to your map-reading with mental images which are all part of the delight and satisfaction of map-reading.

These six skills must be amalgamated if you are to become competent at map-reading, enabling you to face the outdoors with confidence. Beware of spending too much time on aspects such as Grid References at the expense of the practical work out of doors which is outlined in the next chapter.

Navigation: Finding Your Way

The skills of the previous chapter are of vital importance but they are only a preparation. The ability to navigate - to be able to find your way - can only be acquired out of doors by spending many hours relating the map to the countryside. Some basic items of equipment are needed.

Map: The maps listed in Chapter 3.1 - Equipment relate to the maps needed for your ventures. The most effective map with which to learn the skills of finding your way is a 1:25 000 map of the area where you are based or are going to acquire your skills. England, Scotland and Wales are mapped at this scale.

Map Case: The map will need protection if it is not to be prematurely recycled into wood pulp by the rain! A plastic bag will do but, since you will be using the map continually, a proper map case with a lanyard for round the neck will always keep your map at the ready and prevent you from losing it. Map cases are best stiffened with a piece of card or a formica off-cut to make the map easier to use.

Compass: The compass should be of the 'protractor' type with the needle in a liquid filled capsule and a transparent plastic protractor base plate about 10cms or more in length. It is probably not worth paying the extra price for the luminous variety.

Watch: A watch is the most important item of navigational equipment after the map and the compass. Cheap watches today are incredibly accurate, some having a 'stopwatch' facility which is very useful for navigation.

Torch: Remember! In the dark, without a light, a map is about as much use as a sheet of old newspaper. You do not need a lot of light, which will only destroy your 'night-sight', but you may need to use the torch over a considerable period of time. A small torch which will provide enough light to see where you are putting your feet will do, but it must be powered by alkaline cells (no Nicads) and carry spare cells or battery and bulb. A headtorch which leaves both hands free is a luxury and it is also a blessing around the camp site in the Spring and Autumn evenings.

An assortment of simple navigational equipment - some vital, some useful

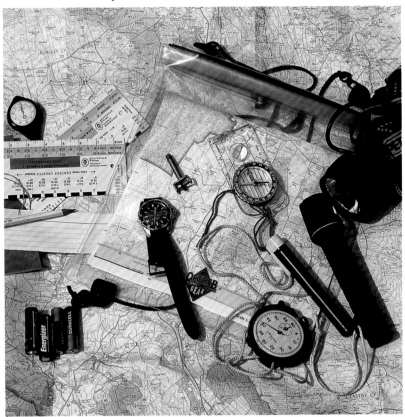

Other items: A Romer and a simple map measurer, or an opisometer, are useful additions, but not essential and the small notebook and pencil carried by all participants completes the list. A soft 2B hexagonal pencil, sharpened at both ends is a most reliable writing implement and ideal for map work.

FINDING YOUR WAY

Finding your way is based on three basic practical skills:
- Locating your position on the map.
- Setting the map.
- Route finding - following a pre-determined route on the ground.

This is made easier if you can:
- Estimate distances visually.
- Know how fast you travel.
- Know how far you have travelled.

If you have a choice, choose a warm day, or dress up warmly, as the ability to navigate is acquired through the use of the head, not the legs, and there will be a lot of standing around comparing map and countryside. The walking comes later! Fold your 1:25 000 map so that the approximate area of your location is in the middle of the map and place the map in the map case.

Locating your Position

Find your home, school or workplace, shop or supermarket on the map and then think out how you located it: by chance, by the names on the map, by searching systematically, by the pattern of roads and buildings? Keep on repeating this process using different landmarks in familiar terrain until it becomes second nature. This ability to locate your position on the map is fundamental to your future success in finding your way. The assistance of an Instructor or someone who can read a map may be most helpful at this initial stage.

Setting the Map

Setting, orienting or orientating the map is the fundamental skill in using the map for route finding. There are two common methods of setting the map by inspection and both should be practised until they are automatic.

By a Line Feature: Look around the landscape and find a line or linear feature such as a road, street, path - anything which has length and direction and is marked on your map. Hold the map horizontal and turn the map round until the feature on the map is parallel to the feature on the ground. The map should now be correctly set, or oriented, with all the landmarks and features in the countryside in the correct direction from the map. This method is simple and effective and you do not need to know exactly where you are providing you have a rough idea. It is possible to have the map exactly the wrong way round, so be careful to check that landmarks are on the correct side of the linear feature you selected.

Fig. 3.6/1

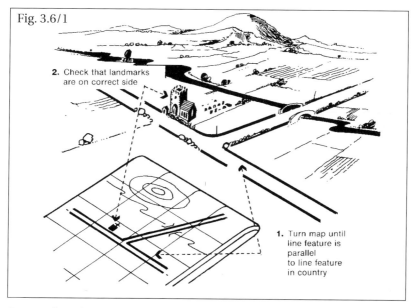

2. Check that landmarks are on correct side

1. Turn map until line feature is parallel to line feature in country

By Landmark or Point Feature: To use this method you must know where you are. Look around the landscape and identify a spot feature which is marked on the map. Lay your pencil or a straight-edge through your position on the map and the spot feature on the map and then, holding the map horizontally, turn the whole map round until you can sight along the straight-edge from your position on the map, through the spot feature on the map, to the feature on the ground. Your map is correctly set. It is a very accurate method when you know exactly where you are, and it will also work quite effectively even if you only know your approximate position, providing you select a landmark which is some considerable distance away.

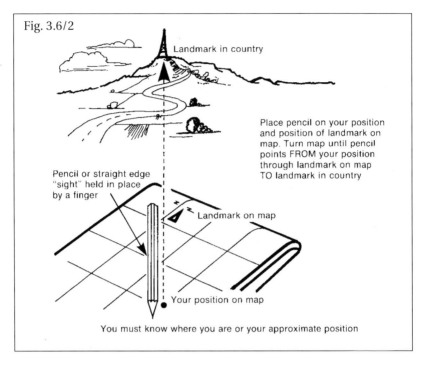

Fig. 3.6/2

Landmark in country

Place pencil on your position and position of landmark on map. Turn map until pencil points FROM your position through landmark on map TO landmark in country

Pencil or straight edge "sight" held in place by a finger

Landmark on map

Your position on map

You must know where you are or your approximate position

Put a map or a street plan in your pocket when you go out and keep on setting the map. Sometimes you will locate your position on the map and then set the map. At other times it will be easier to set the map and then locate your position, but with experience you will find that you are able to carry out both these fundamental processes at the same time. Accustom yourself to looking at the map from all angles so that eventually there is no 'right way round' and to view it from any angle is normal.

A Viewpoint

A map represents an aerial view of the ground which is very different from our viewpoint at ground level. This is something you will have to get used to, but you can speed up the process by finding a vantage point such as a hill, a ridge or even a tall building, which will enable you to look down on the landscape and give the terrain a more map-like appearance. It is worth taking considerable trouble to find a suitable viewpoint which will also enable you to practise the basic skills. When looking over the landscape many people do not know whether a feature is one kilometre or ten kilometres away.

A high viewpoint helps to give the country a more maplike appearance and is especially helpful to the novice

Fig. 3.6/3

Geographical Direction: Once you have set your map you will know which direction is East, South or North East or any of the other geographical directions in the landscape of countryside. East will be the landscape towards the right of your map while North-West will be the country in the direction of the top left-hand corner of the oriented map.

Route Direction: Using the set map 'to point direction of travel' is usually the most important function of a set map. The map is set and a pencil is laid along the path you wish to take or towards your objective. Keeping the map set, lay a pencil through your position along the path you wish to take or towards your objective, hold it in front of you and look along the pencil from your position to your objective. The pencil is now pointing to the direction to follow, or along the path you need to take. This is a most useful skill to use at road, path or track junctions to ensure that you are going to follow the right one. This technique is the basis of route finding with the map.

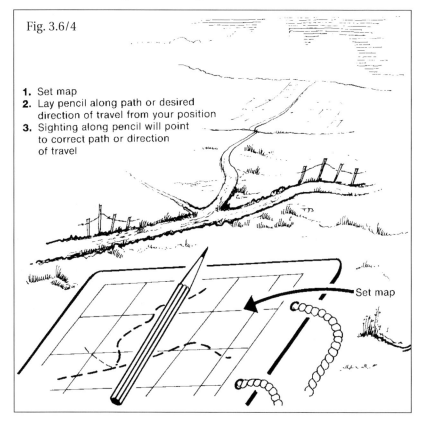

Fig. 3.6/4

1. Set map
2. Lay pencil along path or desired direction of travel from your position
3. Sighting along pencil will point to correct path or direction of travel

Set map

Locating a Feature: Being able to set the map enables you to locate in the landscape a feature or landmark shown on the map. Keeping the map set, lay your pencil through your position on the map and the symbol of the feature you wish to locate in the landscape. Sight along the pencil to the ground and the feature you wish to locate will lie along the line-of-sight. In this technique you are working from **map to ground.**

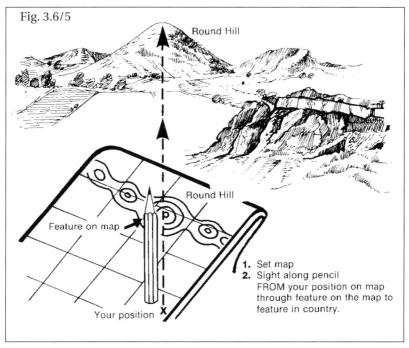

Fig. 3.6/5

Round Hill

Round Hill

Feature on map

Your position **X**

1. Set map
2. Sight along pencil FROM your position on map through feature on the map to feature in country.

Identifying a Feature: This is the opposite technique to the previous one and you work from **ground to map** to name or identify a feature in the landscape. With the map set, place one end of your pencil through your position on the map and then, pivoting the pencil on this point, swing the pencil across the map until it is pointing at the feature on the ground. The feature with its name or symbol should be found on the map under or alongside the pencil, providing it is not too far away and it is marked on the map.

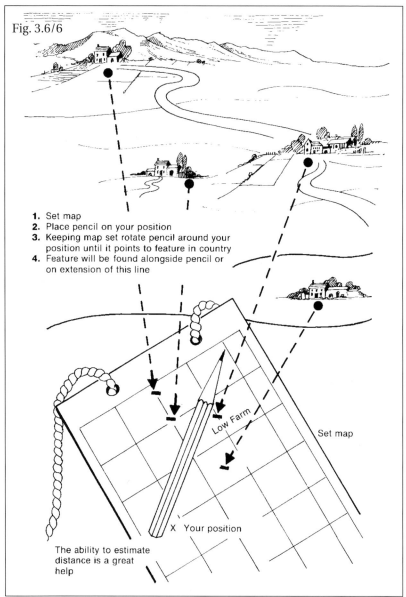

Fig. 3.6/6

1. Set map
2. Place pencil on your position
3. Keeping map set rotate pencil around your position until it points to feature in country
4. Feature will be found alongside pencil or on extension of this line

Low Farm

Set map

X Your position

The ability to estimate distance is a great help

The Visual Estimation of Distance

Some ability to estimate distance by eye is essential for good practical map work and route finding. Some people are better at this than others, but all can improve their ability with a little effort. Go back to your high vantage point and find a prominent landmark and guess how far away it is. Then, using your measuring scale, measure the distance from where you are standing to the landmark on the map. Knowing the correct distance of the feature, look at it again and try and stamp this perception on you mind - the comparative scale of features and any people. If you make a habit of repeating this process using landmarks near, and a considerable distance away, your ability to estimate distances by eye will improve considerably and prevent you from making silly mistakes.

Route Finding

At least 90% of all travel in the outdoors, whether in normal open or wild country takes place on roads, tracks or footpaths because where we go is largely determined by rights-of-way, problems of access and the places from where we can gain access to the countryside. Travelling along a path is also less demanding physically and navigation is easier. A path keeps a traveller on course, as effectively as a railway line keeps a train on its correct route, providing that the traveller is on the right path. Orienteers would call these 'handrails'. The word 'path' will be used in the following text to denote any footpath, track, bridleway, lane or road. Route finding is a two-part process consisting of:

- Selecting a series of paths which will connect your point of departure with your objective or destination.
- Selecting landmarks or features along the paths from the map which will enable you to ensure that you are on the right path and follow your progress along the paths. Landmarks are particularly important at path junctions.

In open or wild country where there are no paths to link the point of departure with the destination, it would be necessary to travel on a compass bearing.

The importance of combining the preparatory skills mentioned at the end of the previous chapter to become 'map-reading' are now apparent, as they enable you to carry out this route finding process.

Using landmarks or features common to both map and ground to identify one's exact position is known as 'pinpointing'. Following one's progress across the countryside by means of pinpointing is usually referred to as 'tracking one's position'. Usually the difference between knowing where you are and being lost boils down to the ability of being able to place your finger on the map and say "I am there!".

Competence comes through practice and it is essential to practise these skills until you can perform them without effort and they become automatic, rather than a tedious chore. Using your 1:25 000 *Pathfinder, Explorer* or *Outdoor Leisure* map, select a series of paths in your neighbourhood which will enable you to travel in a circle back to your starting point. Identify features marked on your map along your chosen route such as path junctions and prominent landmarks which will enable you to track your position along the route. Describe the route to yourself or a friend, using the same method as in the end of Chapter 3.5. Fold the map and place it in the map case. Then walk the route using every opportunity to practise the skills of setting the map, locating your position, locating features, identifying landmarks and estimating distance by eye, followed by checking with the map.

Once the technique has been mastered on the 1:25 000 map, select other routes on a 1:50 000 scale *Landranger* map and perfect the skills at this scale. It will be more difficult as there is less detail shown on the map, but you must be able to work at all scales. The skill of tracking one's position can be practised on a bus, train or from a car but it is probably better to use a 1:50 000 scale map because of the increased speed of travel.

These skills apply to all modes of travel on land, rivers and other inland waters.

TIME, DISTANCE AND SPEED OF TRAVEL

Another very important ability needs to be added to your existing skills - the ability to judge how far you have travelled from your speed of travel, and how long you have been travelling. At first travel will be slow as you will spend so much time comparing map and landscape. This is how it should be! Once you are able to track your movement over the ground while on the move, more attention should be devoted to the speed of travel. A normal walking pace along roads or footpaths might be 5 kilometres an hour, roughly 10-12 minutes per kilometre. Plan a route of several kilometres and, walking purposefully but still tracking your position, time how long it takes to cover sections of the route and the total time of your journey. Repeat this process until you know how long it will take you to walk one kilometre, 500 metres, 250 metres and 100 metres. Once you are able to do this you will be able to work out in advance the approximate time that it will take to complete a journey of a given distance. By adding the estimated time to complete a journey to your time of departure you will be able to forecast an estimated time of arrival (ETA) at your destination. This technique is one of the corner-stones of route planning and a vital aspect of safety in remote areas. **One of the main purposes of practice journeys is to acquire this skill whether you are a canoeist, rower, sailor, cyclist, rider or walker.**

In addition to enabling you to make an estimated time of arrival at a point on your route or destination, it can also help you to navigate more effectively and avoid some of the problems. Progress along a route is tracked by using landmarks, but sometimes there are no obvious landmarks, especially in woodland or rural areas, and it is all too easy, even for experienced map-readers, to miss features, points where paths lead off from roads and tracks, or even path junctions. If you were expecting to find a path leading off from the track after 500 metres and you have not reached it after 25 minutes walking, you must assume that you have passed it without its being identified. One would expect to cover 500 metres in about 8 or 10 minutes at the most. By making a habit of keeping a careful check on travelling time and relating this to the speed of travel you have a vital navigational tool.

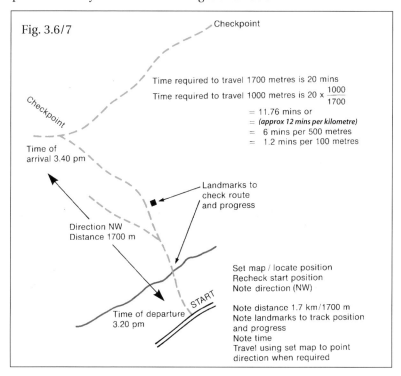

Fig. 3.6/7

Checkpoint

Checkpoint

Time required to travel 1700 metres is 20 mins

Time required to travel 1000 metres is $20 \times \dfrac{1000}{1700}$

= 11.76 mins or
= *(approx 12 mins per kilometre)*
= 6 mins per 500 metres
= 1.2 mins per 100 metres

Time of arrival 3.40 pm

Landmarks to check route and progress

Direction NW
Distance 1700 m

Time of departure 3.20 pm

START

Set map / locate position
Recheck start position
Note direction (NW)

Note distance 1.7 km / 1700 m
Note landmarks to track position and progress
Note time
Travel using set map to point direction when required

Your speed of travel will be affected by your means of travel, the load being carried, the size of the group and the nature of the terrain. These factors balance each other out so it becomes possible, with practice and experience, to make very reliable estimates of the time required to complete a journey.

ALLOWING FOR HEIGHT CLIMBED

Those who travel in hilly country have an important consideration to bear in mind when estimating the time to complete a journey - the extra effort required to travel uphill which is very demanding physically, whether on foot, a bike or a horse. Only experience will enable you to judge what time allowance must be made for climbing, but for those travelling on foot some advice already exists which can be used as a basis for your personal experiments. Various formulae have been devised to estimate travelling time in mountainous or hilly country. The original rule was devised by a Scottish mountaineer called Naismith in the nineteenth century. The rule was intended for mountain walkers engaged in a journey of at least a whole day.

For a fit hillwalker:
- Allow 1 hour for every three miles of horizontal distance to be covered and an additional 30 minutes for every 300 metres or 1,000 feet of ascent.
- Allow 12 minutes per kilometre of horizontal distance and 10 minutes for each 100 metres of ascent (or 1 minute per contour on a 1:50 000 *Landranger* map).

For a fit hillwalker carrying a rucksack with camping gear:
- Allow 1 hour for every 2½ miles of horizontal distance and 1 hour for every 460 metres or 1,500 feet climbed.
- Allow 15 minutes for every kilometre of horizontal distance and 4 minutes for every 30 metres of ascent (4 minutes for every 3 contours on the 1:50 000 map).

Base your own experimentation on these figures and devise your own formula for allowing for height climbed. The importance of practice journeys cannot be over stressed in acquiring this experience. Most young people are not 'fit hillwalkers' and will find the figures very optimistic, especially when it comes to uphill work with a pack full of camping gear. Only very fit Award participants will be able to match the estimates above and the reality is that most groups engaged in Gold

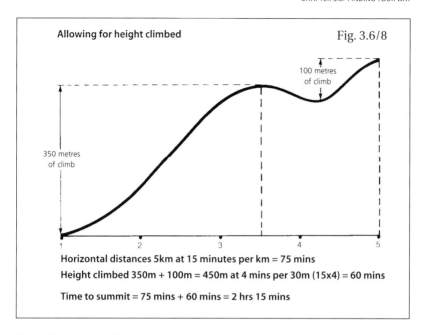

Allowing for height climbed   Fig. 3.6/8

100 metres of climb

350 metres of climb

Horizontal distances 5km at 15 minutes per km = 75 mins

Height climbed 350m + 100m = 450m at 4 mins per 30m (15x4) = 60 mins

Time to summit = 75 mins + 60 mins = 2 hrs 15 mins

Expeditions, travelling through wild country, covering 20 kilometres and limiting the amount of ascent to not more than about 500 metres, will take about 8 hours to complete a day's journey including short breaks.

A SENSE OF DIRECTION

Unlike some creatures we have no in-built sense of direction but we all develop an adequate sense of direction in our childhood for familiar localities. With conscious effort this directional sense can be greatly enhanced and extended to unfamiliar areas. This ability is developed by an awareness of the surrounding environment and observations of natural phenomena. With very few exceptions, all country or terrain has a 'grain' or directional element which can be utilised; this grain can be oriented using the map. By continually orienting the landscape, the landscape can subsequently provide direction for you.

The map of the area in which you are going to carry out your venture each day, or part of a day, should be studied and the boundaries of the area carefully noted. These boundaries are usually delineated by roads, rivers, coasts or ridges and should be unmistakable so that you would realise immediately if you crossed one. If you do get lost, at least you will know the area in which you are lost.

Wind, clouds and the sun all change direction but this change is relatively slow and, in the case of the sun, predictable. These phenomena can all be utilised with careful observation to provide direction and set an alarm bell ringing in the head whenever they are in the 'wrong place'. One of the objectives of the Expeditions Section is to encourage and enjoy a response to the natural world and an environmental awareness.

Navigation: Using the Compass

All navigation has to be based on competent map-reading but there comes a time, when lost, in poor visibility or in the dark, when additional help is needed in the form of the compass. The compass is able to provide direction without reference to external landmarks or features. The modern orienteering compass is so simple to use that it can delay the acquisition of map-reading skills and hinder participants from developing a sense of direction. No attempt should be made to introduce the compass until the skills of map-reading have been mastered.

The compass syllabus is progressive through the three levels of the Award; at Bronze level, the compass syllabus is confined to its care, setting the map by compass and geographical direction in terms of the cardinal and inter-cardinal points. If participants cannot find their way in normal rural country without using a compass then their map-reading skills should be considered inadequate. Exceptions may need to be made where large tracts of forest, chases or woodland, such as the New Forest, are used at Bronze level and Instructors consider it advisable to introduce additional techniques from the Silver and Gold syllabuses. Navigation with a compass must be based on good map-reading, otherwise it becomes a device which, all too frequently, enables the incompetent to lose themselves more effectively. All the groups which get lost within the Award Scheme have at least one compass in their possession!

Many find the total isolation of thick cloud or darkness in remote wild country a frightening experience. Navigation is about mental attitude, having the resolve and the mental toughness to cope with the situation. Participants venturing into wild country will need to go to considerable efforts to ensure that their compass skills are 'grooved' to such an extent that their use is automatic, have been tested by experience, and they are able to cope with the stress which usually derives from such situations. This training will almost certainly involve some work at night and Instructors will have to devise situations where this training can take place in safety and under adequate supervision. If compass skills are acquired on a playing field participants may well find when the real call is made that they have only equipped themselves to find their way round playing fields.

The Compass

The compass should be treated with the respect that a highly sensitive instrument deserves. If it is dropped, the pivot may well be damaged, and, if exposed to excessive heat, a bubble may develop in the capsule. It should be stored away from other compasses, steel, iron, electrical appliances and electrical circuits.

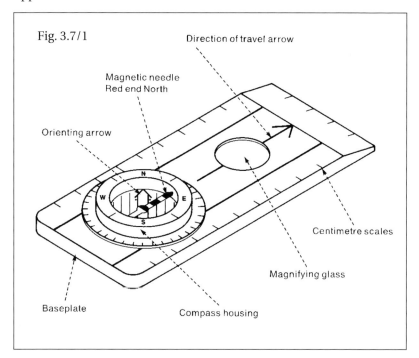

Fig. 3.7/1

Direction of travel arrow

Magnetic needle
Red end North

Orienting arrow

Centimetre scales

Magnifying glass

Baseplate

Compass housing

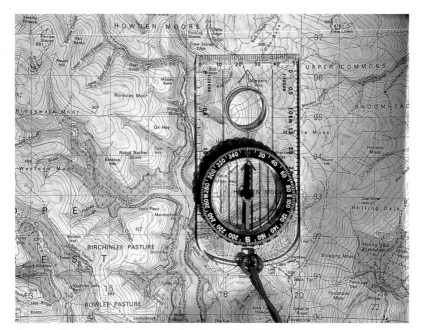

Setting the map by compass - simple and reliable - but try and set it by inspection first!

Do not use the compass close to motor vehicles, metal fences, metal posts or any other large iron or steel objects which, along with power lines, may have a strong electro-magnetic influence and cause the magnetic needle to deviate. Experiment with items of equipment which you carry on your ventures and see what effect they have on your compass. Torches, cameras, watches and penknives are such items and if they affect your compass while in use, they must be packed sufficiently far away to avoid interference.

Setting the Map by Compass

If you are unable to set your map using the two techniques outlined in the previous chapter, it can be set by using the compass. Just place the compass on the horizontal map and turn the map round until the magnetic needle in the compass is parallel with the North/South grid lines on the map and the red North-seeking end is pointing towards the top, or North, of the map. There is no need to bother with magnetic variation as the map will not be more than some five or six degrees out in mainland Britain, less than the error brought about by handling the map out of doors.

In woods and forests it may be necessary to navigate by compass at Bronze level

Geographic Direction

It is very simple to find the principal directions of North, East, South West, North East etc. Hold the compass horizontally with the needle swinging freely and turn the compass, or yourself, around until the needle is parallel with the lines in the bottom of the housing and the North-seeking end of the needle, the red end, is pointing at the 'N' on the housing. The directions of the eight cardinal and inter-cardinal points are now indicated by the housing. If the inter-cardinal points are not marked on the housing, then they can be estimated by eye. South East will be halfway between South and East.

If you wish to travel in one of these geographical directions it may be better to turn the housing round until the desired direction is against the arrow indicating the direction of travel. Then, holding the compass horizontally with the needle moving freely, turn round until the red seeking end of the needle is parallel with the grid lines in the bottom of the housing and the red end of the needle is pointing to the 'N' on the housing. You are now facing in the desired direction with the arrow indicating direction of travel pointing the way you should move. Again, when wishing to travel in these general directions, there is no need to bother with magnetic variation.

The care of the compass, the precautions in its use and the ability to set the map using the compass should be adequate for Bronze level participants in normal rural country. Those using featureless country may have to utilise the additional skills intended for those in wild country.

Checking the Direction of a Path using a Compass

Most of our travelling on land, whether on foot, cycle or horse, takes place on a form of path so it is essential to ensure that we are on the right path or that the path is taking us in the right direction. The competent map-reader can usually check this out using the map alone, but there are occasions when the help of the compass is required.

Paths and tracks wind there way to their destinations so it is essential to be able to see enough of the path to establish its general direction, or the check will need to be carried out a number of times. This technique is particularly valuable at the junctions of paths to ensure that the correct one is selected.

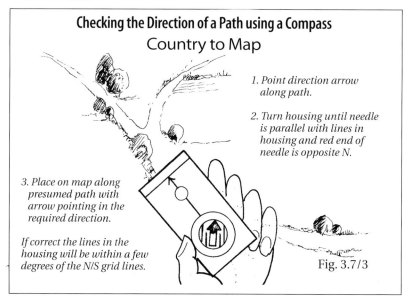

Checking the Direction of a Path using a Compass
Country to Map

1. Point direction arrow along path.

2. Turn housing until needle is parallel with lines in housing and red end of needle is opposite N.

3. Place on map along presumed path with arrow pointing in the required direction.

If correct the lines in the housing will be within a few degrees of the N/S grid lines.

Fig. 3.7/3

Point the direction of travel arrow along the general direction of the path. Holding the compass level with the needle swinging freely, turn the housing round until the needle is parallel with the lines in the bottom of the housing and the red end of the needle is opposite the 'N'. Ignore the needle from now on. Place the compass on the map with one edge along the presumed path and the direction of travel arrow pointing the way you are going. If you are on the right path, the lines in the bottom of the housing will be within five or six degrees of the North/South grid lines on the map. Again, there is no need to bother about magnetic variation in Great Britain or western Europe, it is unlikely that there will be two paths having a direction within five or six degrees of each other.

Fig. 3.7/4

Checking the Direction of a Path using a Compass
Map to Country

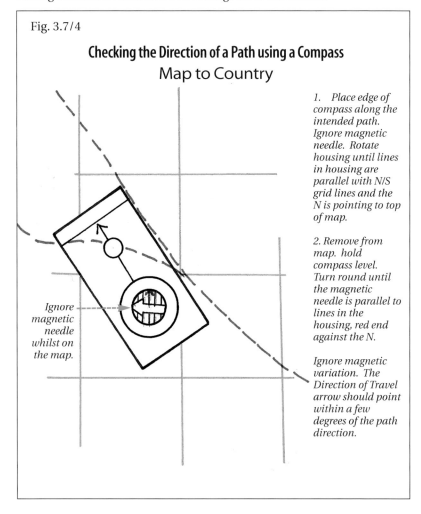

1. Place edge of compass along the intended path. Ignore magnetic needle. Rotate housing until lines in housing are parallel with N/S grid lines and the N is pointing to top of map.

2. Remove from map. hold compass level. Turn round until the magnetic needle is parallel to lines in the housing, red end against the N.

Ignore magnetic variation. The Direction of Travel arrow should point within a few degrees of the path direction.

Ignore magnetic needle whilst on the map.

You can carry out the check the other way round. If you arrive at a path junction and are uncertain which path to take, place the edge of the compass with the top edge along the general direction of the path you wish to take with the direction of travel arrow pointing in the direction you wish to move. Ignore the compass needle. Rotate the compass housing until the lines in the housing are parallel to the North/South grid lines on the map and the North 'N' on the housing is towards the top of the map. You have now obtained the grid bearing of the path. Lift the compass off the map and holding it level and steady, turn yourself around until the needle in the housing is parallel with the lines in the bottom of the housing and with the red end against the 'N'. You and the direction of travel arrow should be pointing along the correct path. Again there is no need to bother with magnetic variation.

Measuring Direction

When using the compass, direction is measured in degrees from the North and always in a clockwise direction.

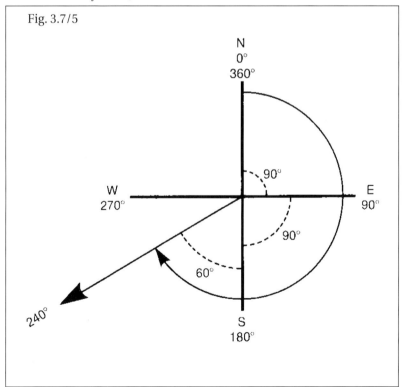

Fig. 3.7/5

Magnetic Variation

The red North-seeking end of the compass points to the earth's North Magnetic Pole, or Magnetic North. This is distinct from the True North, the axis around which the earth rotates. The difference, expressed in degrees, between the True North and the Magnetic North is known as the Magnetic Variation. Sometimes the word 'declination' is used instead of 'variation' but the traditional British use is 'variation'. Magnetic Variation changes from place to place and from year to year and the rate of change is not constant. The difference between Magnetic North and True North is shown in the marginal information on your map. The Magnetic North, depending where you are in the world, can be to the West or the East of True North. In the British Isles the Magnetic Variation is always to the West of True North and will remain so well into the 21st Century. In large areas of the Continent, Magnetic Variation is so small that it can be ignored altogether as it less than the errors which arise in handling the compass and map.

Grid North

In addition to True North and Magnetic North there is a third North - Grid North. Grid North is the northerly direction of the North/South grid lines on the map. The difference between True North and Grid North is so small that, for all practical purposes when using the protractor type compass, it can be ignored. It arises from the difficulty of trying to represent the curved surface of the earth on a flat piece of paper. All directions should be measured from Grid North.

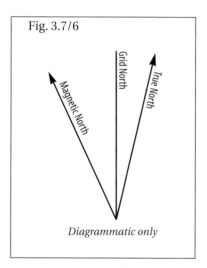

Fig. 3.7/6

Diagrammatic only

Travelling on a Bearing

This is the most vital compass skill, the 'lifeboat drill', of the traveller in difficult situations. It is the skill which enables one to travel from a known location to a destination in featureless terrain or in restricted visibility such as darkness or cloud. It has three steps:

- Using the map to find the grid bearing from your location to destination.

- Converting the grid bearing to a magnetic bearing.
- Travelling on a magnetic bearing.

Each step must be practised until it can be carried out automatically even if your mind is numb with cold or befuddled with anxiety.

Obtaining a Grid Bearing from the Map: Place the compass on the map with the top edge of the compass connecting location with destination and **with the direction of travel arrow pointing the way you wish to go.** Turn the compass housing until the orienting lines on the base are parallel with North/South grid lines on the map and the 'N' on the housing is to the top of the map. The grid bearing can now be read off the scale on the housing opposite the direction of travel arrow. The magnetic needle is ignored completely during this stage.

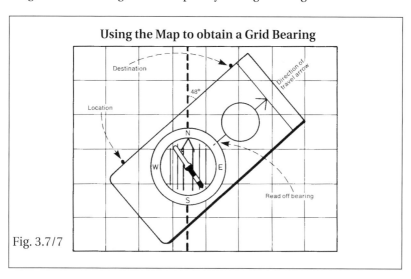

Using the Map to obtain a Grid Bearing

Fig. 3.7/7

Converting the Grid bearing to a Magnetic Bearing: Because the magnetic variation in the British Isles is to the West of True North, the variation is always **added** to the grid bearing. The best way to remember this is that, as the country is larger than the map, the grid bearing should always be made larger when working from the map to the country e.g. using a magnetic variation of 6 degrees West:

Grid bearing	48°
Variation	6°
Magnetic bearing	54°

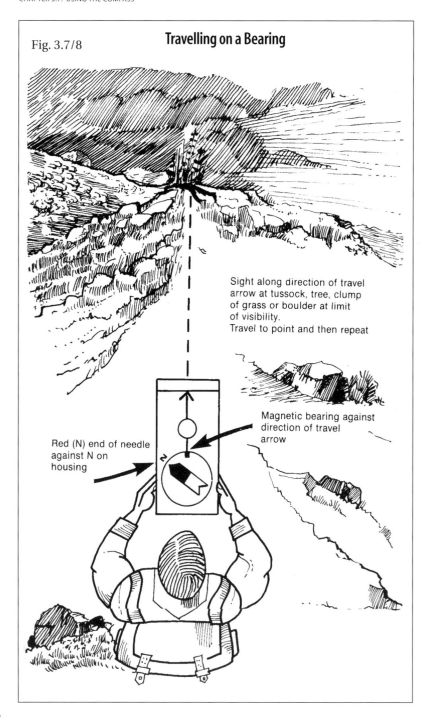

Fig. 3.7/8   **Travelling on a Bearing**

Sight along direction of travel arrow at tussock, tree, clump of grass or boulder at limit of visibility.
Travel to point and then repeat

Magnetic bearing against direction of travel arrow

Red (N) end of needle against N on housing

Immediately after you have found the grid bearing, add the variation by turning the housing anti-clockwise for six degrees.

Travelling on a Bearing: After the bearing has been added, hold the compass horizontally in front of you so that the needle is swinging freely. Then turn yourself round until the red end of the needle is pointing to the North on the compass housing and is parallel to the lines in the bottom of the housing. Sight along the direction of travel arrow at some feature in the landscape: it could be a tree, a clump of grass or a hump on the ground. Fix your eyes on the spot and then travel towards it. When you reach the landmark repeat the process until you reach your destination. The distance between the landmarks on which you sight depends on visibility and the nature of the terrain. In woodland it may only be a few tens of metres, while on an open moor it may well be a kilometre or more.

Although visibility rarely drops below 50 metres in daylight, the compass is usually essential

At night, or in cloud, it may be necessary for a group to provide its own mark to which it must travel. A member of the group, the 'marker', armed with a torch moves forward near to the limit of visibility in the general direction of travel, as indicated by the person with the compass. The 'marker' turns and shines the torch at the person with the compass, who in turn signals the 'marker' to move right or left until he or she is directly on the line of sight. The person using the compass may have to use a torch as well. The rest of the party then moves forward to the 'marker' and the process is repeated until the destination is reached. It can be a slow process but it improves with practice and a forceful, organised approach. This is the simplest method, but for those who are more experienced in the use of the compass, a more elegant method is described in the Award Scheme's publication *Land Navigation - Route Finding with Map and Compass* where the techniques are dealt with in greater detail. During daylight visibility rarely drops below 50 metres except in 'white-outs' and snow storms.

Using the Compass when Lost

It is usually easy to travel on a compass bearing from a known location to a destination but, when lost, the situation becomes much more difficult and the need for positive action more important. Having lost yourself, you must remedy the situation - sitting still and waiting for someone to find you is not an option! Fortunately the situation is rarely as bad as it appears at the first realisation that one is lost.

In England, Wales and Northern Ireland, where the majority of Award ventures take place, it is very difficult to find somewhere which is much more than four and a half miles from a metalled road, though this is not so in Scotland or western Europe. This may not be much consolation in driving rain and a strong headwind, but it does mean that if one follows **a constant direction** it should be possible to reach a road within two or three hours of steady travelling.

The boundaries of areas of wild country are marked out by roads or tracks. These boundaries are known as 'collecting features'. There are some exceptions such as in coastal areas where the sea forms the collecting feature, and rivers and ridges may fulfil the same function. When lost, and all efforts to relocate yourself have failed, the most appropriate action is to use the compass to head for the road or track which surrounds the area. The worst action is to wander around aimlessly in circles.

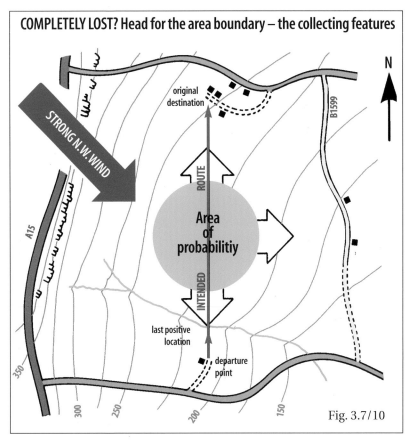

COMPLETELY LOST? Head for the area boundary – the collecting features

Fig. 3.7/10

First note the time. Then, using the map, the time you have been travelling, your speed of travel and the last positive location you made, decide on the area where you are most probably lost - 'the area of probability' - and draw a circle round this on your map. Then decide which part of the boundary surrounding the terrain you are going to aim for. If you have been travelling for a considerable period of time and are confident that for most of the time you have been travelling in the right direction, head for the collecting feature nearest to your original destination. If you haven't a clue, head for the boundary which you consider to be the nearest to the area in which you are lost. Other factors may influence the direction you choose. If the slope of the ground is fairly constant over a large area, you may wish to travel downhill with the slope. There may be a strong headwind with driving snow or rain in a particular direction and it may be preferable to travel in the opposite direction with the wind on one's back. Occasionally the collecting

Gills, gullies and cloughs are wet, slippery and treacherous

Keep to shoulder where there is often a path

feature which is presumed to be nearest may be blocked by an escarpment or a line of steep or difficult terrain which the group may not wish to tackle, especially at the end of the day if there is a possibility of arriving at it in the dark.

Having drawn the ring around the area of probability, place the end of the top edge of your compass in the centre of the area and the other end on the part of the boundary which you wish to aim for. Then carry out the procedure for **walking on a bearing** exactly as if you were travelling from a known location to a known destination. Travel on this bearing until you reach the collecting feature. Do not change your mind unless you are able to relocate yourself without any possibility of doubt. Every effort should be made to increase the standard of map-reading and awareness of the surrounding terrain while carrying out this procedure. In all probability you will encounter paths, tracks and streams which may act as 'handrails' and provide guidance.

Streams can be positive guidance towards lower ground, but in steep terrain they can pose dangers, frequently running in gills or gorges, steep-sided with waterfalls, precipitous drops and treacherous slimy footing. When following a stream in steep terrain, always walk on the shoulders of the stream which frequently have footpaths created by other people who have used them as guides in the past. Because you are lost and only have a rough idea where you are travelling from and only aiming for the boundary of the area - the collecting feature - rather than for a specific place, it may be pointless bothering with magnetic variation and the method used for finding geographical direction will often suffice. Travel towards the most appropriate of the cardinal or inter-cardinal points of East, South West, North etc.

There are three essential compass skills:
*   Setting the map by compass.
*   Checking the direction of a footpath.
*   Walking on a bearing.

These are sufficient for all engaged in ventures in the British Isles and western Europe. The technique to be used when lost is really the same as **travelling on a bearing**. Only when travelling on a bearing is it advisable to compensate for magnetic variation and even this is not necessary in many areas where the magnetic variation is less than the limitations in the construction of the compass and the errors which arise from using the protractor-type compass. There are a number of other useful compass techniques but, unless one uses the compass on a regular basis, they are more likely to lead to confusion, especially in the stress which arises from being lost. Ventures in more distant parts of the world may involve subtracting or adding magnetic variation and the variation may be so large that it cannot be ignored in any of the three techniques listed above without jeopardising the ability to reach your destination.

In normal rural or open country, if the compass has to be used at all, it will probably be for the first two skills of setting the map and checking the direction of paths. Travelling on a bearing is not appropriate for rural country or in the valley bottoms of wild country as it usually involves tramping across fields, meadows and arable land and climbing over fences and walls, incurring the anger of the landowners. Travelling on a bearing is most likely to be used in forest and the upland areas of wild country. The skill may not be used very often but when it is needed and the chips are down, it is vital that you can use the skill automatically, without conscious thought, and still get it right!

Joe Cornish / National Trust

Planning the Route

P lanning the route is only part of the overall planning of the expedition, but it is a very important part and it may play a decisive role in ensuring the success of the venture. Planning a good route is a challenging task which will require enterprise, imagination and a considerable amount of time if it is to be carried out properly. Often it is delayed until the last moment and insufficient time is set aside for this very important task. Once the outline of the route has been agreed, each member of the group should prepare a section, which will still take several hours of combined effort. **For ventures in the Award's designated Wild Country Areas six weeks advance notice must be given in duplicate to the appropriate Panel Secretary on the standard** *Expedition Notification Form* **if an Assessor is required, or four weeks in advance if an Assessor is not required.** A notification number will be allocated to each qualifying venture and should be entered into the participants' *Record Books* on the successful completion of the venture.

Good route planning is essential at all levels of the Expeditions Section and takes on an added significance for ventures in wild country where safety is of paramount importance. Much of the advice which follows is directed primarily towards ventures at Gold level on land, but the principles apply to all levels of Award both on land or water, and should form a sound foundation for all future route planning. Ventures at Bronze and Silver levels may be an end in themselves, but they usually form a progression to Gold level and it is important to establish good practice and habits from the very beginning.

ENSURE THAT THE VENTURE CONFORMS TO THE CONDITIONS

Within the Award Scheme the requirements and conditions will be determined by the level of Award and the mode of travel. These requirements and conditions are clearly set out in the *Award Handbook* which is an essential aid for planning all ventures.

Correct Duration

The planned route must take place over the correct number of days for the level of Award.

In the Right Place

The venture must take place in the appropriate environment as set out in the *Award Handbook* and elsewhere in this *Expedition Guide*. This will be decided by the purpose of the venture and the mode of travel, but all ventures must avoid villages and populated areas unless this is impossible, and for qualifying ventures the country must be unfamiliar.

The Correct Distance or the Required Number of Travelling Hours

For walking and cycling ventures the requirements are in kilometres or miles, while for horse riding ventures and all ventures on water the requirement is measured in hours of travelling time.

ACCESS

Having decided where the venture is going to take place, enquiries must be made to ensure that it is possible to gain access to the desired area or water. This is particularly important for water ventures. The check is best carried out after the locations of the camp sites have been decided as these will almost certainly involve contacting the landowners.

Use of Private Land

Ventures often take place in remote areas where, at first sight, there is little evidence of any agricultural or other activity. All land has some use for its owner and it is important that prior permission should be sought. At certain times of the year, groups may not be welcome for both agricultural and sporting reasons. It is useful to note the dates of the following seasons and remember that deer stalking involves high velocity rifles which can be dangerous at very great distances.

- **Lambing** - *in the valleys around March, April and May.*
- **Grouse Shooting** - *August 12th to December 10th.*
- **Red Deer Stalking** - *July 1st to February 15th.*

There may be other restrictions on access such as Ministry of Defence firing ranges and bombing ranges. The ranges are usually well-defined, certainly as far as paths are concerned, but other access restrictions, such as the boundaries of nature reserves, may be less obvious.

In certain Wild Country Areas there is a tradition of free access to the upland areas with the problems of access being greater in the valley bottoms. The valley floors and sides are enclosed with small fields which provide fodder and winter pasture. They are surrounded by the characteristic stone walls which should never be climbed. The fields are not shown on the 1:50 000 scale map so, for planning purposes, the 1:25 000 scale map should be used as they show the field boundaries and the tracks which provide access to the more open terrain of the higher ground. While rights-of-way are shown on both maps, they are easier to discern on the larger scale map.

Long Distance Footpaths

The use of long distance footpaths for practice or qualifying ventures is not considered to be sufficiently demanding either for planning or navigation. Literature in the form of maps and guides exists in profusion and there are even long distance footpaths associations, all of which deprive participants of the opportunity to plan their own venture. Many paths are waymarked along their length. The sheer volume of walkers using these long distance paths is causing serious erosion, which is having an irreversible effect on the ecology of the countryside and causing great concern. Long distance footpaths should not be used in Award ventures, except in short sections when it is unavoidable, to link up with other paths or rights-of-way.

GENERAL CONSIDERATIONS

The route must be challenging yet within the physical capabilities of each member of the group. There is a tendency to over-estimate fitness and under-estimate the effort required to a carry a pack full of camping gear and food. This is especially so at Gold level where ventures should go **through rather than over wild country - solitude not altitude!** High level traverses of well-known edges, or precipitous ridges and ascents of summits, though naturally tempting, are rarely compatible with the

177

concept of Expeditions and Explorations within the Award Scheme. A distinction must be made between a hillwalker out for the day with sandwiches, some spare clothing and a few items of emergency equipment in a day-sack and those committed to pre-planned journeys through wild country over a number of days, carrying all their camping gear and food, and having to sustain the journey through bad weather when necessary. There is a greater challenge to navigational skills in journeying through mountains using features such as passes and cols, than blindly following parties of hillwalkers along cairned ridges, walks and beaten tracks. The popularity of these 'honey-pot' areas serves to destroy the solitude which can contribute so much to the feeling of achievement which is an important ingredient of successful Expeditions and Explorations. The careful selection of the area and the route is just as important in normal and open country, where the vast majority of ventures within the Award Scheme take place at Bronze and Silver levels.

ADDITIONAL ADVICE

Having ensured that the venture will take place in the right area and the requirements and conditions have been fulfilled, there are other factors which will go a long way to make certain that the venture is successful. This advice is based on the problems and difficulties which have occurred in the past, particularly at Gold level, bringing ventures to an untimely end.

Divide the distance or hours of travelling time equally between the days available.

For a variety of reasons, whether it be the remoteness of the area or getting off work or school, there is a widespread tendency to make a late start on the first day and, in order to return home in good time, the last day is curtailed. Whilst appreciating the difficulties groups may have, this practice results in too much travelling being crammed into the middle days, unreasonable physical demands being made on the participants and, sometimes, unacceptable levels of risk. Excessive mileage on one or two days, frequently with an overweight pack, leads to exhaustion with the accompanying risk of hypothermia in cold weather and exercise-induced heat exhaustion in hot weather. Undesirable side effects occur - groups are too tired to prepare a meal on arrival at the camp site or to make an early start on the following morning. Feet take an additional hammering in foot Expeditions and there is no time to cope with problems if anything goes wrong. The advice given in the

Award Handbook, that participants should spend forty eight hours in the Wild Country Area familiarising themselves with the environment and preparing themselves and their equipment prior to the start of the qualifying venture, is very much concerned with participants' being able to make an early start on the first day of the venture.

Camp Sites

Camp sites are usually located on low land or in valley bottoms for shelter, easier communication and access in case of an accident or an emergency. Groups may lose themselves while journeying but the few incidents, accidents or illnesses which do occur have usually taken place at the camp site.

Travelling Uphill

For ventures on foot or cycle, the minimum distances are mandatory so that the only way to regulate the physical effort is by limiting the amount of climbing each day. While all groups should be concerned with devising challenging ventures, any route which involves more than 550 or 600 metres of climbing in a day should be viewed with suspicion.

Make any Major Ascents Early in the Day

If other considerations permit, plan your route so that the uphill travelling is done early in the day while you are fresh and, if the weather turns nasty later, you will be heading for lower ground and increased shelter from the weather.

Do not Plan Unnatural Routes

With experience it is possible to devise routes following a natural line over the ground which are not only less tiring but may even make navigation easier. Groups follow unnatural routes because they lack experience, or they wish to follow a compass bearing and lack the confidence and the map-reading skills to follow the lie of the land. If you have struggled hard to reach high ground, try and retain this advantage if possible as it is a mistake to lose height and then climb up again, or waste energy struggling up and down gullies, gills, cloughs and steep-sided valleys which frequently abound on the edges of high ground.

179

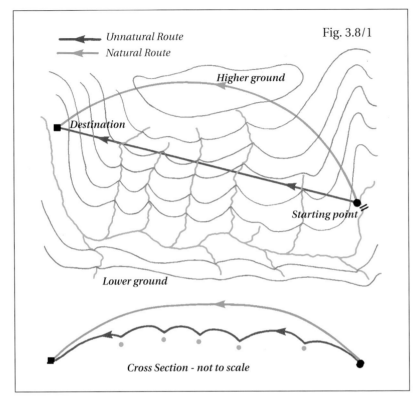

Start Early in the Day

Always start your journey as early as possible in the morning as this will reduce the chances of being overtaken by darkness at the end of the day. This may require a great deal of will-power and self-discipline. Many of the problems which some venture groups have encountered, and the unnecessary calling out of rescue teams, are rooted in a failure to set out at a reasonable time. This is particularly important at the beginning and end of the Expedition season when the hours of daylight are limited. The timing of the Expedition season is as closely related to the beginning and end of British Summer Time as it is to the prevailing weather conditions. Weather does not suddenly improve towards the end of March or deteriorate at the end of October. Groups which start late spend the rest of the day catching up with themselves. Should you have an emergency or get lost, there is more time, if you start early, to sort yourself out and get back on course and, if you are hit by foul weather, you can fall back on your alternative route and still reach your camp site at a reasonable hour. If the situation reaches 'panic stations' at least you will be able to panic in daylight which is better than panicking in the dark - after you have had your panic you will be able to sort yourself out in good time and before being overtaken by darkness.

THE ROUTE CARD

All the considerations listed above have to be taken into account and judgement has to be exercised in balancing out the various, and often conflicting, factors to obtain the best overall route. The route then has to be written down on a route card. The Award has a standard form - a specimen route card is shown on page 183. The route card is an important document as it is a statement of your intentions. While preparing one you have to sort out your ideas, clarify them and commit them to paper. This is a most valuable exercise in itself, but the route card also **informs the Supervisor and the Assessor where you are going and when.** A corner-stone of mountain safety, or safety in wild country, is to tell a responsible person where you are going; the route card does just this! It will also act as a log of your journey. By noting the actual times you arrive at the various checkpoints and then, comparing them with your estimated time of arrival, you will quickly improve your ability to judge how long it takes to travel over different kinds of country. Use a separate route card for each day.

First determine the departure point and the location of your camp site at the end of the day. It is helpful to take a thin piece of string and, using the linear scale on your map, cut off a length equal to the distance you wish to travel each day. Lay this along your proposed route and then make any necessary modifications. Repeat this procedure for each day. After you have planned the day's travel divide the journey into natural divisions of roughly equal length but ending, where possible, at an unmistakable landmark - a checkpoint or a waymark. These divisions are called 'legs', Somewhere between four and eight legs would be a suitable number for a day's journey. The checkpoints may be at a major landmark, such as the top of a col, where the route crosses a road or track or at a major change of direction. Having divided the day's journey into legs using the guidance given in Chapter 3.6 - Finding Your Way, estimate the time it will take the group to travel each leg of the journey, paying particular attention to distance and your own practical experience in judging the speed of travel. **One of the most important functions of practice journeys is to enable groups to measure their speed of travel over similar terrain with reasonable accuracy.**

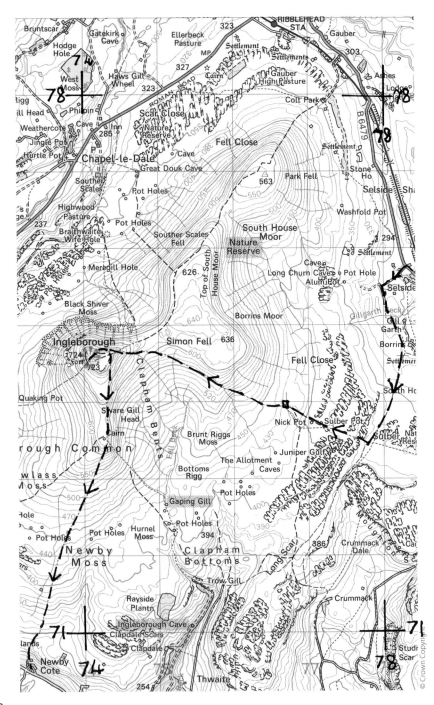

ROUTE CARD (Use one per day)

Day of the week	Date	Day of Venture 1st, 2nd etc
TUES	7th JULY 1996	

NAMES OF GROUP MEMBERS: BARBARA MAIDMENT, MARTYN COX, PAUL RUSSELL, SUE SMITH, ELEANOR HANSON, MILFORD HINDMAN.

NAME OF GROUP OR UNIT: RAILWAY YOUTH CLUB
ADDRESS: RAILWAY APPROACH, HANDOOFT
TEL. No. 019-375-205

Setting out time: 9.00 am.

PLACE WITH GRID REF
START: SELSIDE 783795

Leg	PLACE WITH GRID REF	General Direction or bearing	Distance in km/miles	Height climbed in m/ft	Time allowed for leg	Time for stops or meals	Total time for leg	E.T.A.	Details of route to be followed	Escape to:
1	TO PATH Jct SULBER 777735	SW then S	2·8	80	50	10	60	10.00	Walled path towards Alum Pot then Gillgarth Beck, then path S to Sulber.	Return.
2	TO SHOOTING HUT 766739	W	1·3	50	25	10	35	10.35	Follow path by Sulber and Nick Pots, then by wall.	E to Horton.
3	TO INGLEBOROUGH 741745	W	2·9	300	85	30	1·55	12.30	Path to Allotment Wall then across Simon Fell Breast to summit.	Return.
4	TO LITTLE INGLEBOROUGH 743735	E then S	1·4	–	20	10	30	13.00	Retrace path to Swine Tail then follow path towards Cairn.	Return by same route.
5	TO SWALLOW HOLE GREY WIFE SIKE 737723	between S and SW	1·4	–	20	10	30	13.30	Take SW fork at junction with Gaping Gill path.	South by same path.
6	TO NEWBY COTE 732705	between S and SW	1·9	–	30	1	30	14.00	Follow path or Sike Ladd.	South by same path.
7	TO								ALTERNATIVE ROUTE FOR BAD WEATHER. Follow proposed route to Sulber then path to Clapham and	then path to
8	TO								Trow Gill, then Long Lane to then Old Road to Newby Cote.	93-75205.
	Totals		11.7 km	430m	3hr 50	1hr 10	5 hrs	14.00		

Supervisor's Name, Location, Tel No.: EVAN LITTLE, PETER MILFORD, INGLETON YOUTH HOSTEL, 93-75205.

183

It is a good idea to add a little recovery time, say ten minutes, at the end of each leg. This can be used to 'have a breather' and study the map ready for the next leg. If you are behind your estimated time it will enable you to start the next leg of your journey on time. If you find that you are nearly always behind your estimated time, adjust your speed of travel accordingly and allow yourself more time, and vice versa if you are always arriving too early. If you are having a break for lunch take it at the end of a leg, add sufficient time and remember to allow time for any activity associated with the purpose of your venture.

Finally, add the estimated times needed to cover each leg, plus the duration of the breaks to your intended time of departure each morning to give you an estimated time of arrival (ETA) at your camp site or destination. The departure time each day should be as early as possible or you will be arriving late at all your checkpoints during the day. Do not underestimate the time needed to complete a journey. Groups of young people engaged in Expeditions lasting three or four days progress across the country at about 2.5 kilometres per hour (1½ mph) and a journey of 20 kilometres (12 miles) takes about 8 hours.

Andrew Butler / National Trust

The column headed 'general direction or bearing' is self-explanatory. The general direction of the leg is indicated using the cardinal and inter-cardinal points. Do not enter a compass bearing unless you are able to follow it - they are of no use in normal rural country and when following paths. When you are going to use a compass bearing as a method of navigation, enter the magnetic bearing. A compass bearing may provide an effective escape route in open or wild country, especially if the terrain is featureless. If it is your intention to use a compass bearing as an escape route this, too, must be a magnetic bearing; check that the bearing will not lead you into hazardous terrain.

The 'details of route to be followed' column is concerned with the kind of navigation you are going to use for the leg and will either be in map-reading terms or a compass bearing. Typical examples would be 'follow track, then walk to summit'; 'take path leading SE, then E at jct.'; 'walk on bearing to other side of moor'; 'head E to road'.

To complete the route card it is necessary to include alternative routes and escape routes at the planning stage as a precaution against extreme weather, becoming lost or an emergency.

Alternative Routes

A foul weather alternative route is a route which will enable you to reach your destination and yet avoid the worst of the weather. Usually the route enables you to keep below the cloud base, below the snow line or away from the full force of the elements. An alternative route should be easy to follow, preferably clearly defined tracks, green lanes or old drove roads or, as a last resort, remote country roads which will make navigation easier and can be used with confidence in cloud, hill-fog or darkness. Alternative routes usually add extra kilometres to a journey and involve additional travelling time as they usually 'go round'.

Escape Routes

An escape route, unlike an alternative route, does not usually enable you to reach your destination; it only allows you to escape from the immediate predicament you are in. It may enable you to escape from gale force winds and driving rain to lower and more sheltered terrain, or escape from a ridge in an electrical storm. Escape routes are usually safe ways of descending from high ground and may involve using the compass to the boundary of the area in which you are travelling. You may finish on the opposite side of the hill or in the wrong valley and be compelled to change your plans completely. Your first responsibility on reaching a road or habitation is to try and gain access to a telephone and inform either the Supervisor, the Assessor or another responsible person. Tell them where you are, that you are safe and what your intentions are. You will then have to deal with your problems, sort yourselves out, revise your plans and route, and then endeavour to carry on with the venture. By including intended escape routes on the route card you indicate to the Supervisor and Assessor your most probable course of action in foul weather or in an emergency.

Route Tracings

Finally, after the route card has been completed, along with its alternative routes and escape routes, copies should be made and sent to the Assessor, via the Supervisor, along with the route tracings. While 1:25 000 maps should be used where possible for planning the journey, the tracings should be from 1:50 000 maps as they are smaller and easier for all concerned to handle. For ventures in wild country four copies are needed: one for the group, one for the Supervisor and two for the Secretary of the Wild Country Panel, one of which will be passed on to the Assessor.

Global Positioning Systems (GPS)

For most people, global positioning devices are an interesting curiosity, but it is inevitable that, with the large number of young people involved in the Award Scheme over the next few years, many will find their way into the Expedition Section. It is very important to understand how a GPS instrument can benefit navigators. They are no longer particularly expensive. Just like mobile phones they are becoming smaller, lighter, more user-friendly, more reliable and the power consumption much less, and sales are increasing all the time.

The system works by measuring the time that radio waves, travelling at the speed of light, take to travel from 24 satellites positioned 20,000 kilometres above the earth to the instrument. Knowing the elapsed time and the speed of light, the instrument is able to calculate its distance from the satellite. By calculating the distance from a number of satellites the instrument is able to locate its position anywhere in the world to within plus or minus 25 metres. The system was developed by the United States' Department of Defence for military purposes. For reasons of military security the accuracy for civilian use has been downgraded to a nominal plus or minus 100 metres for 95% of the time. In practice this accuracy may not be achieved, but it is usually possible to locate one's position to within about 60 metres.

The instruments have an integrated electronic compass which will provide an instant course and bearing to any destination. The Silva hand-held GPS instrument can show the right direction immediately without having to travel about 60 – 100 metres. The device provides positions in four ways: as latitude and longitude; in degrees, minutes and seconds; as a national grid reference in England, Scotland, Wales and Northern Ireland; or as a bearing and range from a known reference point thus providing a 'fix' or position. It is a comprehensive navigational instrument able to store a series of way points, course and speed over ground, and estimated time of arrival (ETA) compensation for Magnetic Variation with information displayed in metric, imperial, and nautical unit

Providing the instrument can lock onto four or more satellites it will define altitude to provide a full 3D position. The accuracy for altitude is less than that for a horizontal position, as all the satellites are above the instrument's horizon, and is a nominal plus or minus 150 metres 95% of the time.

The earth shields the instrument from the satellite transmissions. Usually there are sufficient satellites above the instrument's horizon to provide an accurate fix but in gullies, ravines or at the bottom of a cliff the accuracy may be diminished by its inability to lock onto sufficient satellites. Information from three satellites is needed for a fix. The instrument will display the number of satellites from which it is taking information.

GPS are now not much longer than a medium size mobile phone and weigh about 400 grams (less than a pound). The weight and price are reducing and the battery life increasing due to advances in technology. The instruments are becoming more user-friendly, yet at the same time more sophisticated. The trend is set to continue for years to come. Rechargeable batteries should not be used as the voltage will be less and battery life much more unreliable (see Chapter 3.1 – Equipment). The instrument displays the battery state.

The instrument is not intended for micro-navigation. It would be unrealistic to expect an instrument, which is claimed to be accurate to within 60 to 100 metres, to be able to direct you to a ridge, edge or gully leading down from a mountain summit where an accuracy of, say 50 metres, may be required. Such problems are best resolved in the traditional way by pacing on a compass bearing.

GPS is at its most useful when considerable distances have to be covered in darkness, restricted visibility, featureless terrain or extreme weather conditions, thereby avoiding the tedium of packing or the uncertainties of 'dead reckoning' when carried out by the less experienced. Unlike small boats and light aircraft which can usually travel on a bearing, participants in wild country may have to contend with a wide range of obstacles, some of which may present serious hazards.

The instrument will provide a continuous display of bearing, distance and speed over the ground to an objective, but it is vital that the bearing provided is carefully scrutinised on the map to locate any possible hazards well in advance. When obstacles are circumnavigated using the usual orienteering compass, careful pacing and accurate compass work is called for; this is unnecessary with GPS as it will provide a new position after the hazard has been avoided as well as a fresh bearing to the original objective. The full potential of the instrument to store a series of way points enables routes to be planned using natural lines and safe bearings which keep well clear of potential hazards such as precipices or crags.

Additional Skills needed for GPS

All use of the Global Positioning System has to be based on the essential map-reading and compass techniques contained in the navigation syllabus at Gold level as set out in the *Programmes File* and amplified in the *Expedition Guide*. The use of one of these instruments will serve no useful purpose to a person without these skills. It will tell you where you are in terms of a ten figure grid reference or latitude and longitude, but this is not a lot of help unless you can place your finger on the map and say, 'I am here! I need to travel to there!'

Tony French (Halina/Fuji Bursary)

Any participant wishing to use one of these instruments effectively will need additional skills. The simple clarity of a six figure grid reference is all that is needed by a participant at all levels of Award; anything else is a complication of little more than academic interest. Those wishing to use GPS should have a deeper knowledge of the national grid reference system and be able to utilise the marginal information on maps. Using GPS in most of Europe or elsewhere in the world will demand a thorough understanding of defining position by latitude and longitude. You must have the ability to plot positions on a map by hand using scale, pencil and ruler as you may not have the assistance of a national grid similar to that in the British Isles.

An understanding of the principles of traditional navigation and the ability to comprehend and use the basic language and terms of navigation will be of great assistance in developing an understanding of GPS, translating the instructions and using it to its full potential.

Participants in the Award Scheme who are familiar with computers, accessing menus and carrying out their own programming will have little difficulty in finding their way around the display, and inputting data. Studying the instrument and its instructions for a few hours will enable users to move outside and acquire practical experience in its use.

With all items of equipment, especially those concerned with safety and competence, experience in their use is the most important factor. It is usually the user's level of skill which limits performance, rather than the limitations of the equipment. GPS is most likely to be used in earnest when the normal navigational processes have failed, when the group is lost, anxious and stressed, and in extreme weather conditions. It is essential that, in these conditions, the user can feed in the correct data and interpret the information coming out of the instrument accurately.

GPS provides positions and courses. It is essential that the user can plot these and other data on a map. Only then will it be possible for the user to select routes, or escape routes, which serve their need most effectively and are as free from hazards as possible.

Comment

In the hands of an experienced mountain or wild country navigator GPS will provide a most useful additional navigational aid, especially in difficult situations and extreme weather conditions. Unfortunately these devices are more likely to be used as an additional prop by incompetent navigators who hope that a piece of equipment will provide a substitute for competence and experience.

The Award Scheme imposes conditions in the Expeditions Section - ventures must take place between the end of March and the end of October and travel should be through rather than over wild country. All the challenges can be safely achieved with map and compass utilising the syllabus in the *Programmes File* and without the use of any additional equipment.

The global positioning system is in the early stages of development for walkers and backpackers. It is inevitable that GPS will become more accurate as the possibility for this already exists. They will become smaller, relatively less expensive, consume less power and be more user-friendly.

It is likely that the use of GPS will become widespread in the Expeditions Section in the United Kingdom in the coming years. In continental Europe and other parts of the world where the wild country areas are much more extensive and rescue facilities are minimal or non-existent they are being used in increasing numbers.

The use of GPS within the Expeditions Section

The Award Scheme has no recommendations to make to Supervisors and Assessors concerning their own personal use of GPS, but they must not be a substitute for the navigational ability which the Scheme expects of its Supervisors and Assessors. One of the criteria for an Assessor is an intimate familiarity with the Wild Country Area to which they are attached.

Use of GPS by Award Participants

The advice given to participants and their Instructors is basically similar to the advice given on the use of mobile phones. Although GPS does not conflict with the self-reliant nature of ventures to the same extent, their use in wild country is more reliable and predictable.

Where GPS is carried by the participants, the Award Scheme insists that the following procedures should be followed:

1. **There must be no reduction in the syllabus or quality of the navigational training of the group. The Gold level syllabus must be completed in full and the participants must be able to navigate as effectively as a group not using GPS.**
2. **Extra tuition must be provided so that some, if not all, of the potential of GPS can be used effectively.**

This will include:

- Extended coverage of the national grid system, use of ten/eight figure grid references and the map indexing of 1:50 000 and 1:25 000 maps.
- The ability to plot positions, bearings and routes provided by GPS on the relevant maps, and the ability to choose the most suitable routes, alternative routes and escape routes which pose the least hazard.
- Practice in using the instrument's ability to store a series of way points so as to reduce the amount of map work in extreme weather conditions.
- For those involved in ventures abroad, the ability to establish their location using latitude and longitude, plot positions, bearings and routes on foreign maps of the appropriate scale and to communicate such information to Supervisors and Assessors.
- An understanding the terminology of GPS and the ability to ulitise quickly and accurately all the information and data which the instrument is capable of supplying.

- Carrying an ample supply of spare batteries and conserving battery life by judicious use of the instrument.
- Taking care to ensure that the use of GPS does not give rise to a false sense of security. The group must not be dependent on this technology, but must be skilled in the traditional methods of map and compass for sophisticated systems are rarely, if ever, a substitute for 'know-how' and practical experience.
- Assessors and Supervisors ensuring that groups carrying GPS are thoroughly checked on the navigational syllabus as set out in the *Award Handbook*, either at the **Local Pre-expedition Check** or at the **Initial Meeting**, immediately prior to departure on their venture.

Some Instructors may not be able to provide the extra tuition needed to make the fullest use of GPS, especially when they are to be used abroad, and may have to seek help elsewhere. Many Supervisors and Assessors will not be sufficiently familiar with GPS to check the participants' competence with the instrument. They must be content with ensuring that the participants have the fundamental navigational skills and should regard the GPS as an additional skill which the group could manage without in an emergency.

Conclusion

Technology cannot be halted. Technological progress involves every outdoor pursuit and every aspect of equipment from clothing and tents to climbing and caving hardware, and from rucksacks to sailing craft. For many, especially those venturing abroad and participants engaged in Other Adventurous Projects, GPS will, when backed by experience and competence, provide an additional margin of safety but on no account must the absence of GPS, or of a mobile phone, be construed as a lack of safety provision.

Safety and Emergency Procedures

The Expeditions Section of the Award Scheme is all about adventure and challenge, but the very word adventure implies risk. The element of risk has to be reduced to sensible and acceptable levels by training, experience and suitable safety precautions. When we consider the large number of young people who are involved in expeditions every year, it is a great tribute to the commonsense of the participants and the efforts of their Instructors that a mere handful of groups run into difficulties each year. Provided a number of sensible precautions are taken, water and wild country are treated with the respect they deserve, and one remembers that foul weather can cut anyone down to size, the chances of coming to serious harm during a venture are very remote.

Survival is all about mental attitude. Groups have been found shivering, cold and hungry, waiting passively for help to arrive while sitting beside rucksacks packed with tents, sleeping bags and food. If a group gets into difficulties then it is up to the group to get itself out of the situation and not to expect anyone else to extricate them.

Journeying in adverse weather conditions, whether very hot or very cold and windy, can be extremely tiring and it is inevitable that a number of participants, especially those who are not fit, succumb to exhaustion, fatigue, the heat and faint. Such conditions would rarely warrant calling out a rescue team, much less the use of a helicopter. The effective remedy, which can be administered by others members of the group, is rest and plenty to drink - hot or cold as appropriate. Recovery will take place in its own good time and the journey can be resumed or camp set up for the night. There will always be the occasional accident or illness to a member of the group when outside assistance has to be sought straight away.

Again the advice pertains to ventures in wild country and primarily at Gold level, but it forms a base for all venture planning. There are three essential duties to ensure the safety of any self-reliant venture:

Always Tell Someone Where you are Going

Always leave word with a responsible person where you intend to go and the time when you expect to return or reach your destination - your estimated time of arrival (ETA). This procedure is built into the system by providing the Supervisor and Assessor with a route card which always includes the names and addresses of all the members of the group.

Always Keep Together

The group must always keep together. It must not split up or allow anyone to get left behind. When some members of the group are slow this requires enormous self-discipline by the rest of the party. One of the main purposes of practice journeys is to assist in sorting out this problem. Many serious accidents have occurred because of a failure to follow this dictum. The only exception to this rule is when there is an emergency and help has to be sought.

Always Tell the Responsible Person that you have Returned Safely

Participants have a duty to inform the responsible person immediately they reach their destination or of any changes of plans or delay, so that the person will not be concerned for their well-being or initiate any action to find the group.

These precautions not only form the basis of general mountain safety, but apply to all outdoor pursuits where groups are carrying out their activities in some measure of isolation.

Experience over the years in the Expeditions Section has shown that the probability of success and the margins of safety can be enhanced if certain other principles are observed. These are outlined in detail in Chapter 3.8 - Planning the Route. The advice is primarily concerned with the safety of the participants and is repeated below:

- Choose camp sites with relatively easy access in case of an emergency.
- Limit the amount of ascent to avoid unreasonable physical demands.
- Make any major ascents early in the day where possible.
- Do not plan unnatural routes.
- Have an alternative route for foul weather.
- Select possible escape routes in advance.
- Start early in the day.

These considerations, along with an appropriate level of physical fitness and the need to keep pack-weights to the absolute minimum, make a sound foundation for planning ventures within the Expeditions Section.

EMERGENCY PROCEDURES

In the event of an accident or injury first aid must rendered immediately. When an accident occurs it will certainly create alarm and anxiety among the rest of the group. Involvement in the first aid tasks will enable the group to regain their composure before assessing the situation and making decisions. The casualty may be able to struggle on, with assistance from other members of the group, to a place where help or medical assistance can be obtained. The most difficult decision arises when the patient is unable to move. The group may have to split up to enable two participants to fetch help. Before taking this drastic step it may be possible to obtain assistance from other walkers or members of the general public. To attract attention use:

The International Distress Signal

The International Distress Signal consists of six long blasts on a whistle, six shouts or waves of a handkerchief or garment, or six flashes of a torch in quick succession, followed by a pause of about one minute.

The answering signal is: three long blasts, shouts, waves or flashes in quick succession followed by a pause of one minute. Even if you hear an answering signal, continue to send your distress signal until you are sure that you have been located. Your signal will help any rescuers to 'home in' on your position which may be vital in limited visibility. Experienced climbers, walkers and sailors will understand the signal and respond, but ordinary members of the public may not appreciate the significance of the signal, but may come and assist. The probability of attracting attention depends on where you are and when. In certain parts of Wales,

ACCIDENT AND EMERGENCY PROCEDURES FOR PARTICIPANTS

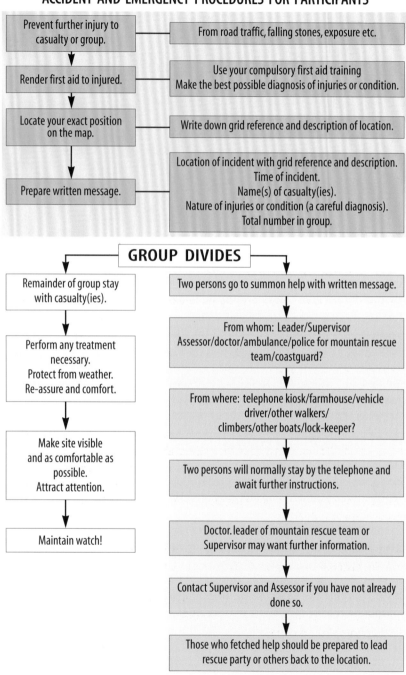

Prevent further injury to casualty or group. — From road traffic, falling stones, exposure etc.

Render first aid to injured. — Use your compulsory first aid training. Make the best possible diagnosis of injuries or condition.

Locate your exact position on the map. — Write down grid reference and description of location.

Prepare written message. — Location of incident with grid reference and description. Time of incident. Name(s) of casualty(ies). Nature of injuries or condition (a careful diagnosis). Total number in group.

GROUP DIVIDES

Remainder of group stay with casualty(ies).

Perform any treatment necessary. Protect from weather. Re-assure and comfort.

Make site visible and as comfortable as possible. Attract attention.

Maintain watch!

Two persons go to summon help with written message.

From whom: Leader/Supervisor Assessor/doctor/ambulance/police for mountain rescue team/coastguard?

From where: telephone kiosk/farmhouse/vehicle driver/other walkers/ climbers/other boats/lock-keeper?

Two persons will normally stay by the telephone and await further instructions.

Doctor. leader of mountain rescue team or Supervisor may want further information.

Contact Supervisor and Assessor if you have not already done so.

Those who fetched help should be prepared to lead rescue party or others back to the location.

the Lake District, Dartmoor or a busy estuary people may be falling over themselves to help you at the weekend. A few miles away, or in the middle of the week, there may not be a soul around to see or hear your distress call. Use your judgement. If you cannot attract attention you will have to help yourselves as there is nothing to be gained from sitting around for the rest of the day blowing your whistle. As a point of interest, in 70% of the incidents in the Lake District which required the involvement of a mountain rescue team, the teams were alerted by members of the party themselves and in 25% of the incidents other walkers and climbers raised the alarm.

Red flares are also a distress signal. White flares are an acknowledgement that the message has been understood.

Fetching Help

No one should dash off to fetch help straight away. Splitting up a group is a drastic step and certain procedures must be followed.

Decide which two people are going to fetch help - usually the best navigator and a fit member of the group. Communal camping gear must be redistributed so that a tent, poles and flysheet are left behind - it is not much use having the tent if one of those fetching help has departed with the poles. Ensure that a stove, fuel and an ample supply of food remains with the injured and companions. Those fetching help should retain their own personal emergency equipment.

The average number of young people in an Expedition group for Gold ventures in wild country is between five and six. A group of this size will enable two people to fetch help and two, or occasionally three, to stay with the casualty. Waiting for assistance to arrive may be a long and anxious time so, where possible, at least two people should stay at the scene of the incident.

The location of the incident must be carefully determined on the map and the grid reference noted. Decide from where to seek help and plan the route. Everyone should know the grid reference, the position and where help is to be sought. A written message must then be prepared. Space for this is provided on the Award Scheme's *Identity Card* which all participants should carry.

It is essential that an accurate diagnosis or assessment is made of any injury, illness, condition or situation before summoning help and the diagnosis or assessment must be written down. Only by being provided with the fullest medical information can a rescue team or the emergency services render effective assistance in the shortest possible time.

Emergency Message

The written message must contain the following information:

- The location of the incident with grid reference and a description of the location.
- The time of the incident.
- The name(s) of the injured.
- The nature of the injuries.
- The number and names of the rest of the group and its origin or identification.

Those fetching help should look around the locality very carefully and memorise the detail, for they may have to lead the rescue party back to the scene of the incident. They should note the time of departure and then set off, looking back at frequent intervals to note the details of the terrain and route from the other direction. They should head for a house, a farm, a road or other feature. In all probability the house or farm will have a telephone. On reaching a road help should be sought from a passing motorist, many of whom may have a mobile phone, or commercial vehicles which will most probably have mobile phones and radios as well. If appropriate, dial 999 and ask for the police as the emergency services are called out by the police. The police, mountain rescue service, coastguard or the ambulance service may wish you to provide additional information. Always carry coins or a phone card in your emergency pack so that you can contact the Supervisor, Assessor or other responsible person and pay for any phone calls. It is impossible to give advice on who should be called first; so much depends on the nature of the incident, the seriousness of any injuries and whether the Supervisor or Assessor can be contacted immediately. Frequently the services of a doctor, obtained through the supervisor, will suffice for illness or medical problems. Remember to thank all those who have helped you!

Waiting for Help to Arrive

The person or people left behind have vital tasks to perform and waiting for help must never be considered as a passive role. The tent will have to

be erected or shelter improvised, the casualty must be kept warm and insulated from the ground and given warm drinks, but **do not give drinks to a casualty likely to require an anaesthetic**. Above all, the patient must be cared for and reassured. The site can be made more visible by attaching a triangular bandage or brightly coloured garment to a stick or wall where it may flutter in the wind. This will help the rescuers who may arrive by land or air.

Frequently the crews of helicopters have difficulty in distinguishing the party needing help because all the other groups in the vicinity wave to them in a friendly fashion. In addition to the International Distress Signal, there is a code of signals for communicating with aircraft.

International Ground to Air Signals
V   Require assistance.
X   Require medical assistance.
N   No or negative. We don't need anything.
Y   Yes or affirmative.

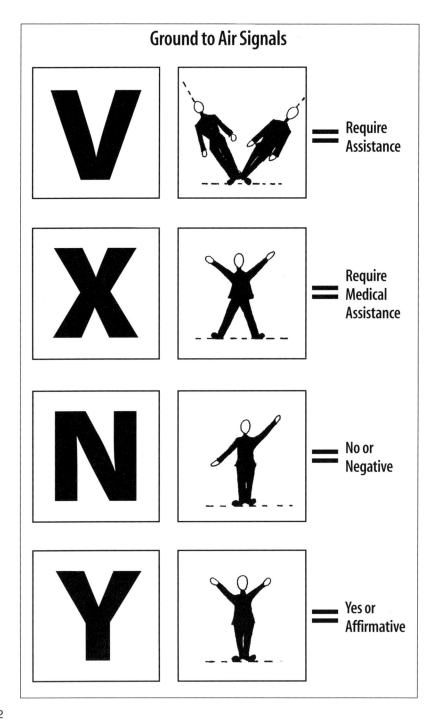

These signals can be made from clothing or anything else to hand and laid on the ground, or by a person taking up the shape of the letter standing up or lying down.

A red flare, a red square of cloth or a fire are also International Alarm Signals in the mountains.

When a helicopter is obviously searching in your vicinity and you do not require help, just carry on with what you are doing and do not wave to the crew no matter how friendly you may feel towards them.

WARNING: At night or in twilight do NOT shine a torch directly at a helicopter for in all probability it will be a Ministry of Defence aircraft and the pilot will be wearing a night vision helmet which increases the intensity of the light to such an enormous extent that the pilot may be blinded.

Help always takes much longer to arrive than people think and it may take several hours. In mountainous areas the people fetching help will have to reach a telephone or ask a motorist for assistance. They will have to dial 999 to call the police who in turn will have to call out the mountain rescue team. The rescue team will have to stop what they are doing at work or home and travel to the rescue post, and collect their equipment before driving to the nearest road-head and setting off to reach the scene of the accident. The time taken for help to arrive usually depends on how far the incident is from the nearest road. Obtaining assistance for water ventures can be just as time-consuming.

It is important for those waiting for help to be alert and ready to help the rescuers to home-in on the casualty. It is helpful if there is more than one person with the injured to keep watch in turns. On seeing help in the vicinity, attract attention by blowing a whistle or waving something bright. At night direct the beam of your torch at a rescue party, but **never** towards a helicopter. This might be the exceptional time when it may be helpful to have a fire ready to be lit if there is fuel to hand.

Emergency Camp or Bivouac

There may be occasions when the whole of the group is forced to bring their journey to a halt. This may be due to weather conditions, being lost, benightment, exhaustion or the first signs of hypothermia, and should not present any great problem to a well-trained and properly equipped group.

In foul weather a tent should be erected in the nearest sheltered spot, as it is vital to keep out of the full force of the wind. Hot drinks should be made followed by a hot meal. If it is impossible to pitch a tent, open it out, place the sleeping mats on the groundsheet and sit on it inside your sleeping bag with all your clothing on and then draw the rest of the tent over the group. Huddle together with legs drawn up under your chin and wrap the flysheet around the whole group.

Do not neglect to make the fullest use of the local environment. Moving position, even a few metres, may provide much greater protection from the wind or sun or could increase the visibility of the location for a search party. A stream may prove to be the solution to heatstroke, or wood and dry vegetation may serve as fuel to provide a means of keeping warm or acting as a beacon. The Award Scheme disapproves of camp fires on its ventures, but there must be exceptions when faced with hazardous situations. Remember the conclusion to Chapter 3.1 - Equipment and utilise the potential of all your equipment to its fullest extent by sharing, pooling and, above all else, improvising.

When the group fails to reach its destination people will become concerned about its safety and a search may be initiated. A conflict of needs takes place. Groups seeking shelter tend to look for hollows, behind outcrops and walls, places which hide them from view, while searchers would understandably prefer a group's location to be clearly visible. Even in the worst weather it is important to make locations more visible and participants have a duty to keep a look out for searchers; whistles and torches should be kept to hand.

Improvisation, the maximum use of all the equipment and resources available coupled with determination, will enable participants to bring about a successful resolution of all their difficulties.

The need to call out rescue teams when groups are overdue is largely rooted in inadequate navigation, lack of physical fitness and overweight packs which make for slow progress. Attention to these factors will go a long way to avoid the necessity to initiate emergency procedures by the emergency services and to give Supervisors and Assessors greater peace of mind.

First Aid

The Award Scheme places great emphasis on first aid and the conditions in the *Award Handbook* stipulate that first aid training may only be given by an Instructor approved by the Operating Authority. This is usually one of the following:

a) An Instructor in first aid recognised by one of the voluntary aid societies, the Armed Services or the Health and Safety Executive.

b) A qualified teacher or youth leader who holds a valid First Aid certificate.

c) A State Registered Nurse or Health Visitor.

First aid procedures and terminology are always undergoing a slow, but steady change and it is vital that participants receive up to date training from Instructors with current qualifications and using the current edition of the Combined First Aid Manual of St. Andrew's Ambulance Association, St. John Ambulance and the British Red Cross Society. This First Aid Manual is essential reading for participants taking part in the Award Scheme. It is detailed and comprehensive and covers all situations likely to be encountered.

The advice in this chapter must always be related to the instruction received during training.

The Instructor in first aid has an important role in preventing two dangerous conditions - hypothermia and heatstroke - by ensuring that all participants are familiar with the early symptoms of these conditions so that they can be recognised at the outset when preventative action will be most effective.

The principles of first aid are the same whether in the hills, on water, the highway or in the home but, on a venture where the group may be remote from immediate help, extra responsibilities will be placed on the first-aider and the patient may suffer more anxiety.

The recommendation is that **each individual should carry his or her own first aid kit** rather than there being communal kit for the whole group. This has the advantage that anyone becoming separated from the group will have a first aid kit to hand. When individual kits are combined, they will provide sufficient resources to deal with major emergencies. An even greater advantage is that participants can customise the kits to their own particular needs. Each kit must include any medicines or treatments which the individual needs for conditions such as asthma or diabetes, together with antihistamines and pain-killers, as it is unlikely that these will be available from any other source. It is the policy that first aid boxes in factories, schools, public service vehicles etc. do not contain medicines or drugs. Participants should know if they have any allergic reactions. An antihistamine may be important and the only effective remedy for dealing with stings and bites. **If participants are allergic to an antibiotic such as penicillin, or any other medicine or drug, they should make this known to the Supervisor, the other members of the group and the Assessor.**

The possibility of encountering the HIV and hepatitis B viruses cannot be ignored. While the probability must not be exaggerated, precautions must always be taken. Disposable plastic gloves are now vital additions to all first aid kits and every effort must be made to avoid contact with body fluids in general and blood in particular. Supervisors and Assessors may wish to carry the more durable latex gloves in their first aid kit.

Many of the problems which afflict those on expeditions are the same as those experienced on holiday or everyday life. They may be annoying and very painful, but they are usually of a minor and temporary nature. The afflictions which are most likely to be encountered on ventures are, in order of frequency:

- Blisters.
- Minor cuts and abrasions.
- Minor burns and scalds.
- Headaches.
- Midge bites.
- Sunburn.
- Splinters.
- Minor sprains.

The first aid kit should contain remedies for these conditions as well as for the more serious injuries which might occur. The contents of a first aid kit are not only limited by weight but by the skill and understanding of the user. The most important items for serious injuries consist of a large, and/or medium, individually wrapped sterile wound dressing, a triangular bandage and a couple of large Melolin squares; the remainder of the items are concerned with making life more comfortable.

First aid kits are best kept in a plastic container similar to those used in kitchens and sold in most supermarkets. Good quality containers are practically airtight and waterproof, and will assist in keeping the contents dry and sterile. Items must be individually wrapped or contained in self-sealing plastic bags to avoid contamination. Tablets sealed in foil are convenient. Recommendations for a suitable personal first aid kit are listed below:

- A large individually wrapped sterile unmedicated wound dressing.
- A medium-sized individually wrapped sterile unmedicated wound dressing.
- An individually wrapped triangular bandage.
- An assortment of individually wrapped sterile adhesive dressings.
- Two or three individually wrapped moist cleaning wipes.
- Melolin squares (or similar) 10 x10 cm or 5 x 5 cm.
- Roller bandage.
- Crêpe bandage.
- Adhesive dressing strip 30 cm x 6 cm for blisters and cuts.
- Chiropody felt or moleskin.
- Pain-killing tablets, probably paracetemol or ibuprofen based.
- Antiseptic cream.
- Anti-midge cream - diethyltoluamide based.
- Sun blocker or **high** factor sunscreen.
- Calamine lotion.

- Zinc oxide plaster.
- Large safety pins.
- Small pair of scissors.
- A pair of tweezers.
- A few pairs of disposable plastic gloves.

The weight and size of the first aid kit, as with all emergency equipment, should be kept to the minimum otherwise the exercise becomes self-defeating. The pooling of resources will usually overcome any individual shortage and the ability to improvise with such items as belts and straps is an important aspect of Expedition training. The weight of the kit should only be 150 grams or so. The mode of travel does not influence first aid treatment or the contents of the first aid kit to any great extent, but it may affect the parts of the body which may require attention. Prevention of injury in water ventures is just as important as it is on land and the wearing of suitable footwear will prevent cuts or stings to the feet. Ventures in southern Europe may increase the possibility of snakebites, although these are still very rare, and heat-related problems may take on an added significance. For ventures outside western Europe, in addition to the personal first aid kit, a larger and more comprehensive communal kit will be needed containing appropriate medicines, drugs and equipment.

Blisters

Blisters are an ever-recurring problem and prevention is better than cure for, within the limitations of an Expedition, it may only be possible to alleviate the condition. The chances of blistering can be reduced by ensuring that footwear is well broken-in before the venture and that socks are free from darns. Care should be taken to ensure that there are no wrinkles in the socks when they are put on. Boots worn with two pairs of socks will cushion the feet and keep them warm but, in hot weather, will soften the feet. In hot, dry conditions walking in trainers with one pair of socks will do much to keep blisters at bay. On arrival at the camp site boots should be removed and, where possible, participants should move around without shoes and socks to allow the feet to harden, but make sure that your feet are not in danger from splinters or cuts or you will only exchange one problem for another.

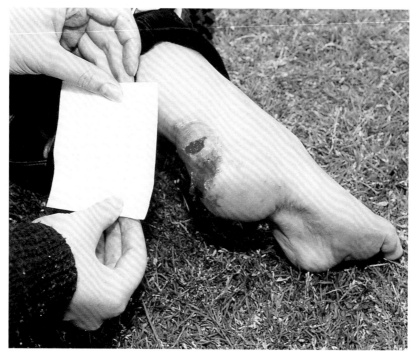

At the first signs of discomfort footwear should be removed, even though it will bring the whole group to a halt. The affected and surrounding area should be covered with thin chiropody felt or moleskin to reduce friction.

Minor Cuts and Abrasions

Minor cuts and abrasions usually require little more than cleaning and an adhesive dressing.

Minor Burns and Scalds

Minor burns and scalds occur while cooking and usually arise from hot pans or the spillage of boiling water. Treatment is by immersion in cold water (or any other cold liquid which is to hand such as milk or lemonade) for at least ten minutes before applying a sterile dressing.

Headaches

Headaches are common, particularly in hot weather. After the journey it is possible to lie down and rest with a cloth on the forehead which has been wrung out in cold water. Otherwise the usual remedy is one or two paracetamol tablets, or your favourite analgesic from your first aid kit.

Midge Bites

Midges can be a scourge on summer evenings, especially in the sheltered places where lightweight campers choose to pitch their tent. A proprietary anti-midge cream containing diethyltoluamide as the active ingredient offers effective prevention.

Exposure to Sun and Heat

In the British Isles sun and heat should not present a serious hazard to fit young people if they take simple precautions. Considerable discomfort and distress can be experienced on Expeditions however, many of which come to a premature end because participants omit to take these precautions. The infrequent really hot summers in this country create far more problems and more ventures are aborted than in the usual British summer. The effects of sun and heat have to be anticipated well in advance as they are more easily prevented than cured. Ventures under the extremely hot, dry conditions common in continental Europe can pose serious problems.

Sunburn

The short-term effects of over-exposure to the sun are so well known that it is surprising that so many participants suffer from sunburn through a failure to cover the skin. Fatigue resulting from headaches and the failure to get a good night's sleep are as significant as carrying a rucksack on burnt shoulders.

Even more serious are the long-term effects. There are around forty thousand new cases of skin cancer every year in the UK, with over two thousand deaths. Skin cancer is one of the most rapidly growing killers of young people in this country. Protection must be provided against the harmful effects of ultraviolet radiation (UVA and UVB). The harmful effects of the third form (UVC) are filtered out by the earth's atmosphere. The amount of ultraviolet radiation is increased with altitude and by reflection from water. Where the skin is exposed, a sun blocker or a high factor (16 or more) sunscreen should be used. This should be water-resistent as this is the only practical form for those taking part in outdoor ventures. New, invisible titanium-based blockers have been developed and these are now available in some chemists.

For groups on the move, especially when there is a breeze, the burning takes place unnoticed until it is too late and the damage done. The body, arms and legs must be protected with loose-fitting light clothing, preferably of cotton which does not impede sweating. Garments made

from new synthetic fibres have been specially designed to give both protection and comfort in hot weather. If shorts are worn, the calves are particularly vulnerable; the legs must be protected at frequent intervals with one of the high factor blocking agents. The head and neck deserve special attention and a hat with a wide, stiff brim provides the best protection. Protection for the neck may be improvised by pinning a handkerchief, a triangular bandage or towel to the back of any hat with safety pins from the first aid kit.

Fainting and Exercise-induced Heat Exhaustion

During the hard, physical efforts of an Expedition in hot conditions, the body temperature can only be kept within safe limits by the process of sweating which leads to an excessive loss of body fluid, this, in turn, places considerable demands on the circulatory system. A frequent and adequate fluid intake must be maintained throughout the day. Salt is lost in sweating and must be replaced. See Chapter 3.3 - Catering for Expeditions.

Fainting is the most common heat disorder and is brought on by fatigue or over-exertion. A short rest, lying with the head down and the legs up, and drinking will usually remedy the situation.

Immobilising Conditions

All the disorders listed above, though potentially serious, have one thing in common: they are unlikely to immobilise the patient for any length of time and, after treatment, all the group will be able to proceed with the venture or seek assistance together. The conditions listed below are likely to immobilise the patient, bring the venture to a standstill, and will probably require seeking outside help. Though these are extremely rare, considering the thousands of participants who carry out their Expeditions each year, all who venture must be prepared for these eventualities. The more serious conditions which may be encountered are:

- Major sprains.
- Dislocations.
- Broken limbs, especially the lower limbs.
- Major burns.
- Injuries to head and back.
- Heat stroke.
- Hypothermia or over-exposure to cold and wet.

Accident Procedures

The procedure after an accident is just the same on an Expedition as it is in everyday life:

- Avoid further injury to the casualty, the rest of the group or the first-aider.
- Maintain airway, maintain or restore breathing and circulation.
- Control bleeding.
- Guard against shock and reassure the casualty.
- Position the casualty correctly but avoid unnecessary movement.
- Treat the casualty to the best of your ability and within the limits of first aid.
- Obtain help.
- Arrange for the casualty to be evacuated.

The treatment of the casualty and the detail of the procedures are set out in the authorised Combined First Aid Manual and must be supported by the mandatory first aid instruction which is a requirement of Award conditions. The first-aider on an Award venture is more likely to be remote from outside help and may have to take action and make decisions which would not normally have to be made in the home, school or workplace, where assistance may be to hand or at the end of a telephone.

Sprains

Sprained ankles happen, even to the most careful walker and the torn tissue swells immediately. If the ankle is covered by the boot, do not remove the boot as this will help to contain the swelling and you may not be able to replace the boot on the swollen foot. Frequently it is sufficient to rest for a while until the pain eases and then it may be possible to proceed, assisted by a stick or a shoulder to lean on. If there is a stream or water nearby immersing the foot, complete with boot, may help to reduce the swelling and ease the pain. If the ankle is not covered, then a crêpe bandage, soaked in cold water and tied in a figure of eight over the shoe and ankle will help to reduce the swelling and provide a little support. Occasionally the damage to the surrounding tissue is extensive and you may not be able to distinguish between a sprain and a fracture. In such cases there is no alternative but to treat the injury as a fracture and outside help may have to be sought to evacuate the casualty.

Major Burns

The most likely cause of major burns arises from the improper use of stoves. These accidents will most probably occur while camping. The most important action, after preventing further burning, is to douse the affected area with large quantities of cold liquid; this should not present a problem at the camp site which will probably be located near a water supply. This cooling must continue for at least ten minutes and will help to reduce pain. While this treatment is taking place, arrangements to evacuate the casualty to hospital must be initiated. Do not touch or remove any clothing which is sticking to the burn or apply any ointments. Try to prevent infection of the wound by improvising some form of clean, dry covering such as a triangular bandage or even a plastic bag or cling-film, although this will not be as easy as at home or in the workplace. The patient must be monitored and the usual checks of airway, breathing, circulation and shock must be carried out.

Broken Limbs

Broken arms and wrists occur less frequently than broken legs. It may be possible when the injury is to the arm that, after treatment and immobilisation of the fracture, the casualty will become 'walking-wounded' and reach a place where medical assistance can be obtained. The very rare head, back and chest injuries will, almost certainly, require experienced outside assistance. **If a spinal injury is suspected, expert help must be sought and the casualty must not be moved unless absolutely necessary.** The Award Scheme's advice to travel through rather than over wild country, thereby avoiding exposed ridges and edges while carrying a full rucksack, should prevent the very serious injuries to head, spine and chest which may result from a fall.

Heatstroke

Heatstroke is a more serious disorder which may occur during Expeditions in hot conditions, particularly to unfit participants, when the body becomes dangerously overheated. The victim feels dizzy, giddy or faint with a rapid pulse rate, often followed by a headache, vomiting, muscle cramp, fainting and collapse. Treatment involves resting the patient in a head-down position in the shade, natural or improvised, and cooling the head and body. Cold water should be poured slowly over the clothing on the upper part of the body using a water bottle or cooking utensils. The clothing will help to retain the moisture and promote evaporation, making a more effective use of the water if it is in short supply. A folded wet towel should be placed on the forehead. The patient should be fanned with a map case or towel. If water is readily available, sponging the skin with an improvised sponge will be very effective. Alternatively, immersion in a nearby water will relieve the condition. Sipping plenty of cold water or other drink, to which a pinch of salt has been added, will replace body fluid and speed recovery. The treatment must be continued until the situation has been completely stabilised.

If, like hypothermia, the symptoms are recognised in the early stages and prompt and effective action is taken, recovery is usually rapid. Heatstroke is a dangerous condition and medical assistance should be sought as soon as possible. If the patient becomes unconscious, breathing and pulse should be carefully monitored and resuscitation administered if necessary.

Keeping the head and body well covered, frequent drinks, well-salted food and a rearranged schedule to enable travelling to take place in the early morning or the cool of the evening, instead of during the heat of the day, is a simple recipe for successful Expeditions in hot weather.

D.W.Elson

Treatment for heatstroke

Hypothermia

Hypothermia, or exposure as it is frequently called, happens when the temperature of the core - the interior of the body containing the vital organs such as the heart, lungs and brain - falls below 35 degrees C. It is a dangerous condition and if the temperature continues to fall, unconsciousness, breathing and heart-failure leading to death will result. Individuals vary considerably in their resistance to cold, but the young and the old are particularly vulnerable to hypothermia, as are those who fall into cold water which removes heat from the body thirty times more quickly than cold air. The probability of hypothermia is therefore probably greater in water-based activities than on land.

Hypothermia occurs when the body is exposed to cold, whether brought about by falling into water or the cold, wet windy conditions of the British hills. The British Isles pose a far greater danger than the more predictable hard, dry cold of Alpine regions. Wet destroys the insulation provided by clothing and the wind lowers the effective temperature drastically - the wind-chill effect.

The other critical factor in hypothermia is exhaustion. The draining of the body's reserves of energy by the physical demands placed upon it is nearly always a major factor in hypothermia which occurs under Expedition conditions. **Continuing the venture, or trying to increase the work rate to keep warm, only succeeds in draining away what little energy remains and makes the condition worse.**

Signs and Symptoms

It is not always easy to distinguish between someone who is very cold, a frequent and common condition for all who engage in outdoor pursuits, and the first signs of hypothermia.

When conditions are really bad, members of a group must watch each other's condition and behaviour to detect the first signs of hypothermia. The well-being of individuals may entirely depend on the observations of the rest of the group. It is part of the nature of hypothermia that it cannot be recognised by the victim. The following signs and symptoms may occur, though not in any particular order:

- Complaints of feeling cold, miserable and tired.
- Pale skin and shivering.
- Lack of interest in what is going on and a failure to understand simple questions and directions.
- Violent shivering ('the shakes').
- Lack of muscular co-ordination, slurred speech and abnormality of vision.
- Irrational, aggressive or violent behaviour.
- Unusually cold skin to the touch.
- Shivering stopped, a slowing down of the pulse and breathing.
- Collapse, coma, death.

Treatment

If the signs of hypothermia are detected at the very earliest stage, it may be possible to seek shelter or beat a hasty retreat while the potential victim can still function effectively. Urgent action is required. Sometimes hypothermia can affect a single member of a group while the rest are still fit. This can occur when an individual is less fit than the rest of the party, is exhausted, has a medical condition, is particularly thin or is recovering from an illness such as 'flu. The initial signs of hypothermia in a participant should ring an alarm for all the other members of the group who may themselves be not far removed from the hypothermic condition.

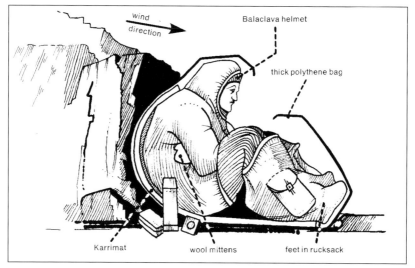

The procedure is outlined for ventures on land because the greatest number of Expeditions take place on foot or by bicycle, but the routine is the same for ventures on water. Hypothermia can readily occur after a capsize in cold weather. For canoeing, rowing and sailing ventures on rivers and canals and confined to inland waters where the bank can easily be reached, treatment should not present a problem; neither should therapy pose a problem on keelboats. There can be a real problem, though, for those in canoeing or dinghy sailing ventures on the sea, in estuaries or other coastal waters, unless a safety boat is present. Hypothermia is most likely to occur in exposed places where there is no shelter from the wind and where the wind blows strongly and relentlessly. These conditions are found on hills, mountains, moorlands, rivers, in estuaries and on the coast.

The first action is to find shelter from the wind in the lee of a wall, rock or ridge. Further heat-loss must be prevented, the victim being covered with extra dry clothing and placed in a sleeping bag and then inside a bivvy bag. Cold strikes up from the ground so insulation must be provided using a sleeping mat or anything else which can be improvised. If the victim's clothing is wet and there is dry clothing available, the wet clothing should be replaced as quickly as possible, keeping heat-loss to a minimum. If there is no dry clothing, the patient should be placed, wet clothing and all, into a bivvy bag and then into sleeping bag. This will avoid the sleeping bag getting wet and losing its insulating qualities.

While the casualty is being insulated, shelter should be provided, either by pitching a tent or constructing a bivouac, and water boiled for a hot drink. If the patient is conscious and able to swallow, hot sweet drinks will warm internally and replace lost energy. This can be followed in due course by boiled sweets and chocolate bars which provide energy quickly in an easily absorbed form.

Warmth will be increased if the group huddles together and it may be possible to get another person into the sleeping bag or bivvy sack to keep the patient warm.

No attempt should be made to warm the victim's skin by rubbing or using a hot water bottle as this may cause a surge of blood from the core to the surface with fatal results. Do not give alcohol.

Normally recovery will take place of its own accord by slow re-warming in a matter of hours, or sometimes more quickly.

If the patient stops breathing, resuscitation should be carried out. The pulse of a victim in the advanced stages of hypothermia may be so weak that it can hardly be detected and may only be found in the carotid artery in the neck. **No effort should be spared to detect any pulse as any attempt to administer cardiac massage while the heart is still beating, no matter how weakly, may have fatal results.** A casualty, especially one suffering from hypothermia, must not be presumed dead by anyone other than a doctor. 'Persons who are cold and dead must not be presumed dead until they are warm and dead!'

Unless hypothermia has been caught at an early stage, especially if collapse and unconsciousness have occurred, assistance must be sought to evacuate and treat the casualty as it requires the competence of those who are experienced and qualified in such treatment. Hospitals, paramedics and mountain rescue teams have more sophisticated methods of re-warming and are capable of providing careful, gentle handling and horizontal carrying which are essential in advanced hypothermia.

In conclusion, hypothermia should never happen to those engaged in Award ventures during the Expedition season, whether on land or water, as participants will be carrying all the equipment required to prevent it - food, warmth and heat in the form of a stove and fuel, insulation in extra clothing, sleeping bags and mats and the shelter of a tent. **Providing everyone in the party is alert to the insidious way in which hypothermia creeps upon its victims in cold, wet, windy conditions and all keep a constant watch on each other for the first symptoms of hypothermia, it should not pose a threat for, like the effects of sun and heat, it is more easily prevented than cured.**

Weather

Weather is beyond our control so it is essential that we learn to live with it wherever we may be. Our modern life-style shields us from the weather and many millions of city and town dwellers never experience the full impact of the weather during their lifetime. This is also true, to a lesser extent, of people who live in rural lowland areas. For many young people, venturing into the mountains or coastal waters for the first time can be a very enlightening experience. Weather is by far the most important factor outside our control which will determine the success, or otherwise, of our ventures. The weather is always more extreme in upland and exposed areas. Winds are stronger, temperatures lower and the rainfall greater, and our safety and success may well depend on the ability to cope with these conditions. Suitable clothing and equipment is vital in shielding us from the elements, but there comes a time when the weather is so bad or the visibility so poor that it is impossible to proceed with a venture. It becomes necessary to escape the full force of the weather by retreating to lower ground and implementing an alternative route plan, or even postponing the venture until a later date.

It is essential, for both success and safety, to have an up to date weather forecast before setting out; this takes on an added significance for those who venture on water. Weather forecasting has improved out of all recognition in recent years with the introduction satellite photography and radar. Occasionally the timing will be a little out and it is still impossible to predict whether you will be under one of the scattered showers or basking in the sunshine only a mile away.

William Fediw

The ability to make local weather predictions for the immediate future is important

It is not possible for Award participants to make a forecast as accurately as the Meteorological Office. Participants should, however, be able to:

- **Make the best possible use of weather forecasts on television, radio, and telephone.**
- **Know how, where and when to obtain a weather forecast.**
- **Make a local weather prediction for the immediate future.**
- **Anticipate weather changes.**

To provide a comprehensive insight into the weather is beyond the scope of this *Expedition Guide* and there are a number of excellent books on this fascinating subject. Form a habit of watching weather forecasts on the television as they probably provide the easiest way of gaining an understanding of such terms as depression, cold fronts, warm fronts and anti-cyclones. On the following page are some simple generalisations which may help those engaged in Expeditions in the British Isles during the Spring, Summer and Autumn.

To make the most effective use of weather forecasts, it is useful to know the elements which combine to make up our weather. The important factors in weather are:

- **Temperature**.
- **Rainfall**, or more correctly, precipitation as it may also fall as snow or sleet.
- **Wind** with its two important components: **speed** and **direction**.
- **Cloud**, with its three components: **amount, kind and height**.
- **Visibility**.
- **Air Pressure**.
- **Humidity**.

Temperature and Rainfall

Normally, temperature and rainfall are the two most important factors in our daily life determining whether we will be warm or cold, wet or dry. Carrying out a venture in extremes of temperature, whether hot or cold, may make travel very difficult unless one has the equipment, skills and experience to cope with them. Rain poses problems because it destroys much of the insulation provided by dry clothing, but this can usually be overcome by the use of protective overclothing. Water ventures may have to be aborted as a result of too much or too little rain.

Wind

Wind is more significant in Award ventures than in daily life, especially in exposed areas, and may well become the dominating factor. The first factor, speed or strength, exacerbates the effects of rain and low temperatures. Strong winds give rise to driving rain which can penetrate clothing and equipment. Walking, cycling or canoeing into driving rain, sleet or snow can be a daunting experience and plays havoc with planning. Wind lowers the effective temperature drastically - the wind-chill effect. A moderate breeze of 24 km per hour (15 mph) at 0° Celsius (32° Fahrenheit) would feel very cold and have the same effect on the human body as a still air temperature of -10° Celsius (15° Fahrenheit). Wind speed increases with height and augments the wind-chill factor in the hills, while too much or too little wind may prevent a sailing venture taking place at all.

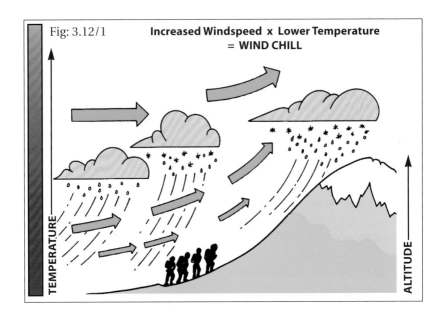

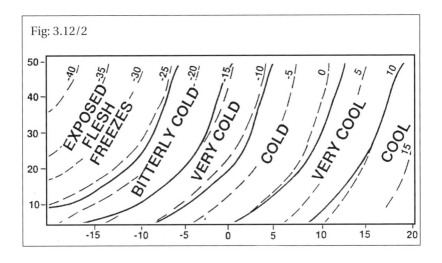

© *Crown copyright reproduced with the permission of the Meteorological Office*

The direction of the wind may affect our weather over a considerable period of time; wind brings weather. Wind direction is determined by the geographical direction from which it comes. Meteorologists give names to these air streams, such as Polar Maritime or Tropical Continental, and the same winds may bring different weather in Summer and Winter. It is sufficient for participants to associate certain weather conditions, or temperatures, with wind direction. The prevailing wind in the British Isles ranges from the North West to the South West. It is usually cool from the North West and mild from the South West, frequently bringing rain and associated with depressions or 'lows'. Changes in wind direction usually lead to changes in the weather.

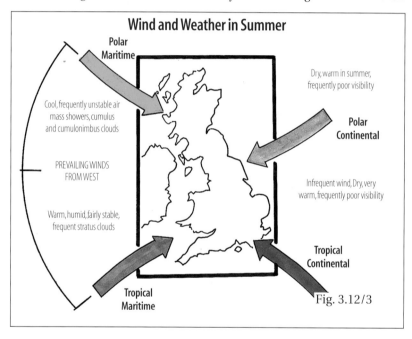

Wind and Weather in Summer

Polar Maritime

Dry, warm in summer, frequently poor visibility

Cool, frequently unstable air mass showers, cumulus and cumulonimbus clouds

Polar Continental

PREVAILING WINDS FROM WEST

Infrequent wind, Dry, very warm, frequently poor visibility

Warm, humid, fairly stable, frequent stratus clouds

Tropical Maritime

Tropical Continental

Fig. 3.12/3

Cloud Coverage

Cloud cover is measured by how many eighths of the sky are visible, where 0 indicates a completely clear sky and 8 represents a totally overcast sky. It prevents the formation of fog and mist in the morning and evening, especially in the Spring and Autumn. A cloud covered sky acts as a blanket, reducing the radiation of heat from the earth's surface which keeps the temperature up during the night with no frost or dew, and little condensation in the tent. A cloudless overnight sky, especially in the Spring and Autumn usually results in a cold night dew, or even frost, and condensation within the tent. This is when you hope that you have not camped in the local 'frost hollow'. Blanket cloud usually means warmer camping, though thick and extensive cloud cover frequently brings heavy, prolonged rain resulting in wet camping and travelling.

Cloud Types

Clouds have been used in weather forcasting for hundreds, if not thousands, of years. The different kinds of clouds and their descriptions used by meteorologists would fill many pages of this book, but they are based on four cloud forms and their height: cirrus, cumulus, stratus and nimbus. Participants may become aware of one or two familiar cloud forms. Fair-weather cumulus - the small, white fluffy clouds of the morning which increase steadily in size during the day and then rapidly diminish and disappear after sunset. Cumulonimbus - the very large and deep thunder clouds, often with their accompanying anvil heads and the layers of stratus which frequently cover the sky.

The cloudbase may determine visibility on high ground

William Fediw

Cloud Height

Clouds are classified as being of high, medium or low height. Continuous layers of low cloud (stratus) frequently sit on the tops of mountains, moorland and upland areas shrouding the tops and summits in cloud or hill fog. This is very easy to observe and it may be of considerable significance to those on Expeditions. The bottom of the layer of cloud is known as the **cloud base** and a competent map-reader will be able to establish the height of the cloud base from the contours on the map. If the participants' route takes them above the cloud base they will be navigating in cloud or hill fog, and visibility may be restricted. These layers of blanket cloud tend to persist at the same height over a considerable period of time. A rising cloud base tends to indicate that the weather is going to improve, while a lowering cloud base usually means that the weather is closing in. The tops of hills, mountains and moorlands are regularly covered in cloud in both Summer and Winter.

Visibility

The information on visibility included in weather forecasts is directed towards sailors, pilots and drivers and usually has little significance for those engaged in Award ventures, except for those venturing in upland areas where cloud or hill fog may make route finding very difficult. During daylight hours, hill fog or low cloud rarely reduces visibility to less than 50 metres unless it is accompanied by driving snow. The early morning mists and fogs of lowland areas in the Spring and Autumn are soon burnt away by the sun and rarely present problems to participants at Bronze or Silver levels when finding their way along bridleways, footpaths, rivers or canals. Venturers in estuaries or coastal waters may, however, be faced with problems arising from coastal fog or sea-frets.

Air Pressure

Air pressure is of great importance to the professional weather forecaster and many interested in the weather have a barometer in their hall. It has less relevance to those on expeditions as it is unlikely that they will be carrying a pocket barometer. Many dedicated hillwalkers and climbers invest in a pocket barometer as it is most useful for making weather forecasts and may assist navigation. A falling barometer, a backing wind (wind shifting in an anti-clockwise direction) and a sequence of cloud types with a gradual lowering of the cloud base provide a very reliable way of predicting an oncoming depression with its consequences. Rising pressure usually indicates a tendency towards better weather while a falling barometer may indicate bad weather.

Humidity

It would be most unusual for participants to be able to measure humidity while on a venture, though pocket hygrometers do exist. The effects of high humidity are very obvious making physical exertion a sweaty, sticky process in the muggy atmosphere. High humidity coupled with heat plays a most significant role in the formation of thunderstorms, which is not a very favourable environment for physical exertion as it usually leads to increased fatigue.

OBTAINING A WEATHER FORECAST

There are three main sources for obtaining weather forecasts:

- Television.
- Radio.
- Telephone.

Television forecasts probably provide the best information, being in a visual form and indicating the general situation and weather patterns, with an expert to provide an interpretation. There is one problem; it is most unlikely that you will encounter many television sets during the venture! Some newspapers also provide a weather forecast but they are no more likely to be available than a television and not as up to date. Every effort should be made to watch a weather forecast, especially on the BBC, for several days prior to departure. This will enable you to build up an understanding of the general weather situation and weather patterns. It may be possible to gain access to a television forecast during the acclimatisation period. Forecasts look forward for several days and even a week ahead.

The radio, especially BBC Radio 4, provides comprehensive information in addition to specific forecasts for sailors and farmers. The myriad of local radio stations frequently provide weather forecasts and weather warnings for the locality. Make a list of the times of the principal weather forecasts on both television and radio and keep it in your map case or pocket. Forecasts need to be up to date so, if you cannot watch or listen to one yourself, get someone else to write them down or record them for you.

It is possible to obtain weather forecasts by telephone. A network of sources with the relevant telephone numbers cover the country and it is possible to obtain a forecast for a particular area, often for a week ahead.

The National Parks Centres may also provide weather forecasts for the users of the parks. Forecasts are usually pre-recorded so you will need to write down the essential information. A little research in the daily press and telephone directories will be amply rewarded as these forecasts are probably the most readily available for venturers. The relevant telephone numbers should be added to the list of the times of television and radio forecasts.

MAKING LOCAL WEATHER PREDICTIONS FOR THE IMMEDIATE FUTURE

Weather conditions are hardest to predict for mountainous and upland areas; mountains make their own weather. Weather forecasts must always be supported by your own observations. All ventures can be made more comfortable and pleasant by anticipating what is going to happen weatherwise, 'keeping a weather eye open', as the sailor would say. When you see heavy dark clouds upwind of you then it is reasonable to expect rain, especially if you can see 'fall streaks' under them. This is the time to don waterproof overclothing or get your tent down while it is still dry, or put the tent up in a hurry if you have just arrived at your camp site. A lowering cloud base engulfing the 'tops' may indicate the need for careful navigation before being engulfed in cloud.

Heavy or Prolonged Rain

Thunderstorms, when accompanied by torrential rain, or just very heavy rain, frequently give rise to flash flooding when streams, becks and burns which can normally be hopped over without a second thought, are quickly turned in raging torrents. Such is the weight and force of moving water that this presents a very serious danger. It is usually very difficult, or even impossible to stand in such water when it is around knee height - an Award Expedition when participants are carrying full backpacks is not the time to experiment or find out. Streams and rivers in spate should be avoided, even if this involves a long detour. If the stream near to your camp site is swollen or in spate after overnight rain, then you may assume that all the other streams in the vicinity will be in a similar condition, and it is time to examine your intended route very carefully to see if it is going to present serious problems. It may be necessary to re-route part of your day's journey and make more use of bridges. Prolonged steady rain over a day or two may bring about the same conditions, but you will have more time to adjust to the situation.

William Fediw

Heavy or prolonged rain turns streams into raging torrents

Thunderstorms

Thunder and lightning can occur at any time of the year and at any time of the day or night. The Summer thunderstorm, with its accompanying lightning strikes, is fairly easy to predict; it is the product of heat and humidity. In central and southern Europe, especially in mountainous regions, they can be a daily occurrence over a period of a week or more. They are characterised by large or huge cumulonimbus clouds extending to heights of many thousands of metres and covering very large areas. The cloud may be capped by an anvil shaped cloud when viewed from a distance. They build up during the day and these thunderstorms are more likely to occur towards the end of the day or during the early part of the night rather than in the morning.

It is difficult to obtain statistical information on the number of people struck by lightning; one thousand a year is probably about right for the United States of America. The vast majority of those struck are engaged in outdoor pursuits, with golfers and climbers apparently the most vulnerable. Many adopt a philosophical attitude to lightning believing that nothing can be done, so why worry! This is not so; it is possible to take very effective steps to reduce the probability of being struck by lightning.

The anatomy of thunderstorms and lightning is extremely complex and it is sufficient to say that large charges of static electricity build up in 'cells' in thunderclouds, some carrying a negative charge while others carry a positive charge. Lightning occurs when these enormous amounts of static electricity are discharged. About half the lightning strikes take place within the clouds and do not reach the ground at all. These strikes appear to those on the ground as 'sheet lightning', though it is exactly the same as 'forked lightning'. Negative charges collect at the base of thunderclouds and create an opposite, positive charge in the ground underneath. The discharge of millions of volts between cloud and earth gives rise to a lightning strike - the 'forked lightning'. A leader strike is usually directed towards the highest ground under the cloud, or to prominent objects on the ground such as trees, buildings or you and me. This is followed by a return stroke from the earth to the cloud. Thunder, which many find frightening, is nothing more than the noise made by the explosive expansion of the air by the electrical discharge.

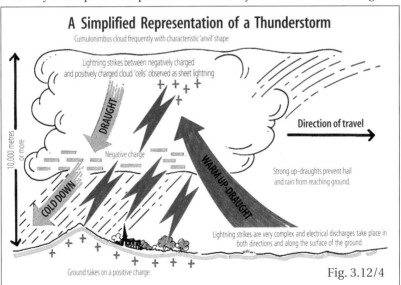

A Simplified Representation of a Thunderstorm

Cumulonimbus cloud frequently with characteristic 'anvil' shape

Lightning strikes between negatively charged and positively charged cloud 'cells' observed as sheet lightning

DRAUGHT

Direction of travel

10,000 metres or more

Negative charge

WARM UP-DRAUGHT

Strong up-draughts prevent hail and rain from reaching ground.

COLD DOWN

Lightning strikes are very complex and electrical discharges take place in both directions and along the surface of the ground.

Ground takes on a positive charge

Fig. 3.12/4

The distant roll of thunder and the cumulonimbus cloud, with or without its anvil, warns of an approaching thunderstorm and the need to retreat to lower ground because summits, high ridges and bare mountains are not good places to be in thunderstorms. Keep well away from trees, especially isolated ones for, even if you do not receive a direct hit from the lightning, the tree could be explosively reduced to woodchips which may be as lethal as bullets. If you are in a forested area try to move to a clearing or an area where the trees are not as tall or are on lower ground. If you are out in the open, head for the lowest ground or a depression and sit, but do not lie, down. Do not be afraid of getting wet for the chances of being struck by lightning may actually diminish in heavy rain.

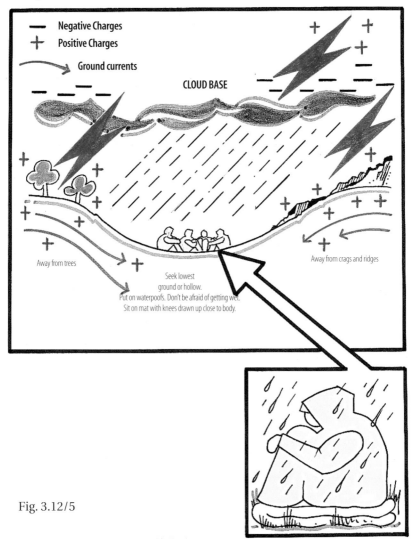

Fig. 3.12/5

Put on your waterproof clothing and then sit on your sleeping mat, or rucksack, with your feet tucked as closely under your body as you can, with hands folded in front of you. This will help to prevent any of your vital organs being damaged by lethal ground currents which arise close to lightning strikes. Lightning strikes are not usually instantly fatal, though the victim will almost certainly be unconscious and may be severely burned. Lightning can strike the same place twice, but this is most unlikely, especially if you have followed the advice above or are in relatively low or open terrain. Normal resuscitation procedures can be carried out on the spot and, if successful, the usual treatment for any burns can follow, before evacuation to hospital.

If you are on the water, on a river or other inland water and there is an approaching thunderstorm, make for the bank or shore as you will be standing out like a sore thumb on the water. Then follow the advice outlined above. If you are in a dinghy or keelboat and you cannot reach the shore, then there is not a lot that you can do about it. The author, when faced with an electrical storm and unable to reach the shore, fastened the anchor chain (even the smallest anchors usually have a few metres of chain attached) tightly to one of the wire shrouds and dangled the rest into the water. This formed a continuous metal conductor from masthead to water. The effectiveness of the action taken was not put to the test, but at least it inspired confidence at the time. In a yacht, all the expensive electrical and electronic instruments should be disconnected.

It is possible to tell how far away a thunderstorm is, or whether it is approaching or receding, by counting the seconds between the flash of lightning and the thunder. Sound travels at about 3 km per second, or 5 seconds per mile. A delay of 15 seconds would indicate that the storm was 5 kilometres or 3 miles away. Thunderclouds suck in air in advance of their direction of travel so they frequently appear to approach against the wind. Thunder and lightning associated with the cold front of a depression can occur at any time of the year, but it is usually no more than a distant roll of thunder in the British Isles.

Finally, if you are at base camp, or have access to a motor vehicle, climb inside as the metal body work will act as a protective shell around you - a 'Faraday Cage' - giving complete protection.

Foul Weather

There are references in certain chapters, such as Chapter 3.8 - Planning the Route, to alternative routes for use in foul weather. It is difficult to provide a satisfactory definition of foul weather. By implication it is worse than bad weather which is a normal experience, especially in open country, upland areas or wild county and, with its nautical overtones, it may be easier to define for coastal waters. Those who sail in coastal or off-shore waters will probably have a good idea of what is meant by foul weather.

Finding a definition is difficult because there are so many factors involved: the mode of travel, the location of the venture and the competence and experience of the participants coupled with their familiarity with foul weather, as well as the suitability and the effectiveness of their equipment.

Probably the best example of foul weather might be the weather associated with a deep depression - gale force winds, driving rain, low cloud and, in the Spring, sleet and snow, especially on higher ground. This would correlate well with the few ventures which come to a premature end and the limited use that is made of alternative routes and escape routes. The two most significant elements in foul weather are high winds and limited visibility. Strong to gale force winds on open or coastal waters do not need any explanation. On land, gale force winds mean penetrating rain, an increased wind-chill effect and difficult, exhausting progress into a headwind. Camp sites have to be chosen with care or they may be flattened. Confident, skilled and experienced groups are prepared to tackle low cloud on high ground, especially when they are using old pack-horse roads, miners' tracks, drove roads and green lanes, but steep and difficult terrain present serious problems in bad visibility and the majority of groups find the isolation and the lack of orientation very disturbing. Probably the best working definition of foul weather has to be 'weather conditions which seriously jeopardise the participants' ability to arrive safely at their destination'.

The British have always talked about the weather, probably because of its unpredictability compared with that in the rest of Europe. Expeditions provide the opportunity to turn the talking into a greater understanding through experience and observation, especially when it is coupled with reading one or more of the many excellent books on meteorology and weather forecasting.

Fitness for Expeditions

A dried up river-bed can bring a canoeing Expedition to an end as effectively as the absence of wind may stop a sailing venture, or gale force winds and torrential rain may bring a journey on foot to a premature finish. The weather is beyond our control and we have to accept its infinite variety as stoically as we can. It is a great tribute to the tenacity and determination of the participants within the Award Scheme that they overcome adversity so frequently.

While we cannot do anything about the weather conditions, it is a matter of great regret that many ventures come to a premature end as a result of problems created by the participants themselves; they are victims of their own shortcomings. Studies over the years reveal that the most frequent reasons for ventures being aborted arise from three inadequacies:

- Poor navigation leading to groups losing themselves and becoming exhausted by having to travel extra miles to rectify the situation.
- Overweight rucksacks, particularly in walking ventures.
- Lack of physical fitness.

Of these three factors, a lack of physical fitness is the most prevalent shortcoming and the most difficult to overcome. Even the best groups get lost occasionally, but one competent navigator may resolve the group's problems and, providing the group is fit, they can cope with any extra distance involved in sorting themselves out. Bathroom scales and the ruthless elimination of unnecessary items can help to reduce the weight of a rucksack. The loads can be redistributed, or items may be discarded en route when meeting the Supervisor. Should the weight of the load be excessive, fit individuals are better able to manage. It is the combination of overweight packs, together with the additional problem of travelling uphill and a lack of fitness which proves to be the disastrous combination. Just as one very competent navigator can salvage an Expedition, so one unfit member of the group can jeopardise the success of the venture.

It is essential that physical fitness is considered at the initial planning stage when participants form their groups, and at every subsequent stage during their preparations for the qualifying venture. Participants should try to find companions who have a similar level of fitness as their own, or who are prepared to train to reach a satisfactory standard of physical competence. Should the rest of the group be fitter than you, you will hold them back; if on the other hand, your companions are much less fit than you, or are not prepared to spend the time achieving physical fitness, then you will become frustrated.

One of the most significant actions participants can take to ensure the success of their venture is to form a group where all the participants are prepared to agree some form of contract with each other to train and achieve the necessary levels of fitness.

In a society centred around mechanical transport and television, fitness is now a matter of national concern. It is becoming increasingly difficult for participants to carry out Expeditions of 80 kms (50 miles), although this is exactly the same distance as when the Award Scheme was introduced forty years ago.

Due to the national concern with health and fitness, the British Medical Association and the National Health Authority are recommending 30 minutes of light exercise per day in its advertising campaigns. The recommendation is for brisk 30 minutes of walking the dog, cycling, gardening or any activity which will result in a slight breathlessness. This is a recommendation for adults - such exercise will do little for the fitness of those undertaking Expeditions. If all the participants in the Expeditions Section devoted 30 minutes a day, or three and a half hours a week, to more strenuous exercise, then practically all the problems associated with a lack of fitness would go away. There would be a greater expectation of success, less frustration and, above all, much greater enjoyment on the part of many individuals. Travelling would give greater pleasure and, instead of participants focusing inwardly on their bodily condition, they would be able to enjoy, to a greater extent, the pleasure of travelling through the countryside.

By making the exercise more strenuous it is possible to reduce the amount of time to three or even two and half hours a week divided into three equal sessions. The work-rate should be gradually increased until fitness and stamina is built up. This will involve making yourself breathless and increasing your heart rate, though increased fitness will result in a slower pulse rate at rest.

The Timing of Fitness Training Programmes

Ideally all participants should aim to maintain a level of training and fitness throughout the year. When this is not possible, the timing of the training becomes important. As a rough-and-ready guide, if you spend two or three months getting fit, then it takes around two or three months to lose this fitness. If only the minimum amount of time is going to be spent on training, then this should commence about two or three months before the venture and should become increasingly strenuous until two or three days before the venture. There is little merit in exercising hard for a month in April to achieve a high standard of fitness and then forgetting all about it until the qualifying venture at the end of July; it would be better to commence training in the middle of June.

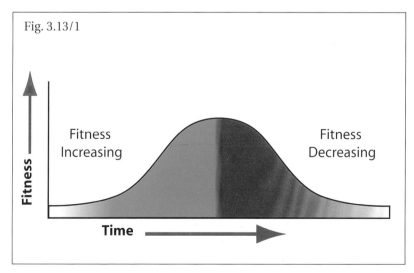

Fig. 3.13/1

Someone who is unfit can ruin the group's prospects of success and spoil the enjoyment of all concerned by becoming an endless and frustrating drag on the rest of the group. We have often seen a group member lagging a few hundred metres behind, the rest of the group stopping and waiting dutifully for the person to catch up. When the person catches up, the group sets off again at their normal pace and the process is repeated which is demoralising for the person who is not fit and frustrating for those who are! A training committment is necessary from each member of the group. It is more satisfactory if this training can be a group activity rather than carried out on an individual basis as members provide each other with mutual support and encouragement, especially in bad weather.

A Level of Physical Fitness

How fit does one have to be to complete a venture? This is difficult to quantify but the 12 minute fitness test below will enable participants to make a more accurate assessment of their level of fitness and embark on a training programme. The level of fitness is determined by a very simple test - how far can you travel in 12 minutes by running, jogging or walking? This is based on the researches of K. H. Cooper MD, MPH, who introduced the world to the concept of aerobic exercises and popularised them in the sixties. To determine the level of fitness requires no more than a watch with a second hand or a stop-watch and some level, or fairly level, ground which has been measured out to a mile or 1500 metres. It could be the perimeter of a football or hockey pitch, part of a playing field or street pavements which would have the added advantage of being illuminated at night but, wherever it is, the environment should be safe and secure. Improvement in fitness is brought about by following a regular training programme and measuring progress using the twelve minute test, details of which are given below. The exercises used for achieving fitness are running, swimming, cycling and walking, probably in that order, and their appropriateness for the Expeditions Section is clear. With the possible exception of swimming, all the training can take place out of doors. There is great merit for those embarking on an Award venture to train out of doors in the wind and rain.

The Twelve Minute Test

The origins of this test are American and it is therefore given in imperial measurements, but it is easy to metricate.

The 12 Minute Test

Distance Covered	Level of Fitness
Less than 1 mile	Very poor
1 - 1¹/₄ miles	Poor
1¹/₄ - 1¹/₂ miles	Fair
1¹/₂ - 1³/₄ miles	Good
1³/₄ miles or more	Excellent

Alternatively fitness can be expressed as the time taken to travel one mile or 1500 metres.

One mile		1500 metres
over 12 minutes	Very poor	over 11 mins 11 sec
9 min 36 sec - 12 min	Poor	8 min 56 sec - 11 min 11 sec
8 mins - 9 min 36 sec	Fair	7 min 27 sec - 8 min 56 sec
6 mins 51 sec - 8 min	Good	6 min 23 sec - 7 min 27sec
Less than 6 min 51 sec	Excellent	Under 6 min 23 sec

These are intended as a guide. 1500 metres is 120 yards short of a mile. Participants would therefore be less tired covering 1500 metres than they would be covering a mile. It will be necessary to clip a few seconds off the metric distance as these have only been reduced proportionately. The times are for young men - young women may increase these times by 5%-10%.

General Fitness Training

Fitness training should take place on at least three days a week. At the end of the training session participants should be out of breath, tired and sweaty. The old adage 'if it isn't hurting, it isn't working!' applies to fitness training. Once the first few training sessions have been completed, most participants will find considerable satisfaction and experience a great feeling of well-being and relaxation. Running and jogging are very popular, but hard swimming, cycling and walking, or any mixture of these, will do equally well. Every week a timed trial over the measured mile or 1500 metres should take place and the time recorded in a log. In this way participants will be able to measure their progress up the fitness scale using the table above.

Any form of aerobic exercise which brings about a strengthening of the cardiovascular system will be beneficial to participants in the Expeditions Section, but it is essential that the exercise becomes progressively more strenuous. A training programme should be devised and agreed by the group and be used as a yardstick to measure individual and group progress. Getting fit is not complicated or involved, it just requires effort, but certain precautions must be taken:

- Exercise must always be proceeded by a gradual warming up period.
- There should be a cooling down period after the exercise.
- Suitable clothing should be worn. Where running or jogging takes place on the road, highly visible clothing should be worn with prominent reflective stripes to make you more visible in poor light, at dusk or in the dark.

Stamina is a vital component in all ventures in the Expeditions Section and the emphasis must be on cardiovascular and respiratory training and endurance. The Physical Recreation Section described in the *Award Handbook* will provide additional information, especially programme on

Upper body strength is helpful in water ventures

Keith Spillett

Fitness Activities, and participants should consider following one of these programmes. It is essential that the activity from the Physical Recreation Section should be timed to coincide with the training demands during the preparation for the venture so that physical fitness does not 'peak' too early or too late.

Specific Fitness Training

General fitness training is necessary to improve stamina but there is a great difference in the demands which the various modes of travel make on the body. Ventures by cycle and on foot make considerable demands on the lower part of the body, while rowing and canoeing require strength in the upper part. Where there is access to a gymnasium, it may be possible to seek advice and devise training programmes, such as weight-training exercises, which will deal with these needs. Swimming will provide the endurance and strength for the upper part of the body needed for water ventures in addition to providing greater familiarity and increased levels of safety in the water. Upper-body strength, combined with strengthening those muscles which assist lifting and handling is a great advantage in the loading of canoes, boats and dinghies on and off trailers, up and down river banks and at portages.

Gymnasia and fitness centres are not essential for training. Many believe that the best training for hillwalking is hillwalking, the best training for canoeing is canoeing. Many mountaineers prepare for their ventures by walking to work carrying a heavy pack - usually filled with bricks. Getting up earlier in the morning and walking to work, the office or school and back again in the evening is a good recipe for fitness. Regardless of the general fitness programme or specific fitness training, it is essential that all participants should prepare for their venture by practising their mode of travel in a progressively strenuous way.

For ventures on foot, this training can be concentrated by walking uphill with an increasingly heavy pack, the maximum being a little above a third of the body-weight, and gradually building up the speed of travel. Cycling up and downhill with heavy panniers and increasing the distance and pace will fulfil the needs of the cyclist. Similarly, progressive programmes for canoeing, rowing and sailing should be followed using heavily loaded vessels and gradually increasing the speed of travel.

Training which involves using the intended mode of travel has other advantages. Feet are prone to blisters in walking ventures and hands may suffer in canoeing and rowing. Training which involves the actual mode of travel:

- Draws attention to those areas of the body which are subject to hard wear and damage.
- Hardens the affected parts and enables participants to learn how to prevent these problems.
- Gives participants the opportunity to learn how to treat a problem when damage does occur.

Any training programme for the Expeditions Section which does not attempt to improve physical fitness is an inadequate one. More ventures end through lack of fitness than through lack of technical skills. Ensure that you and your companions are as fit as possible and you will then be able to look forward to the venture with confidence, knowing that you will be better able to face all that the weather can throw at you. Of even greater importance is the satisfaction which will arise from knowing that you are on top of your task and the journey will be remembered for the right reasons, as an enjoyable and exhilarating experience.

Caring for the Countryside

C are for the environment in which ventures take place is so important that it must permeate all the activities involved. Much of the advice on care and conservation is given in the relevant chapters concerned with the mode of travel and the skills and activities common to all ventures.

It is a requirement that all ventures take place in normal rural, open or wild country, or on water. One of the objectives of the Expeditions Section is to enable young people to enjoy and appreciate the countryside and accept the consequent responsibilities for its care. The Country Code, set out below, is designed and forms a foundation to help the public to care for the countryside and all who live and work there. If this code was implemented by all who use the countryside for recreation, then the country would be cleaner and tidier, but it only forms the basis of care, a 'bottom line', and all Award participants should have much greater awareness and a more responsive and sensitive attitude to the environment.

The Country Code

- Enjoy the countryside and respect its life and work.
- Guard against all risk of fire.
- Fasten all gates.
- Keep your dogs under close control.
- Keep to rights-of-way across farmland.
- Use gates and stiles to cross fences, hedges and walls.

- Leave livestock, crops and machinery alone.
- Take your litter home.
- Help to keep all water clean.
- Protect wildlife, plants and trees.
- Take special care on country roads.
- Make no unnecessary noise.

These twelve carefully worded and selected statements form the foundation of our care of the countryside. The greatest impact made by participants on the countryside is most likely to arise from walking, camping and cycling.

Walking

The modern lightweight walking boots and trainers specifically designed for walking have far less impact on the ground than the heavier boots of the previous generation, but they still cause serious erosion. Large tracts of our long distance footpaths and other popular paths are very seriously eroded and the Countryside Commission, National Parks, the National Trust and Local Authorities spend very large sums of money repairing them. The conditions of the Expeditions Section preclude the use of long distance footpaths, except for short distances to connect areas of wild country or other rights-of-way when it is unavoidable. This helps to mitigate the effect of erosion. Where participants encounter a section of footpath which has been repaired or artificially surfaced, they must confine their walking to the preserved path and on no account walk on either side.

Groups should not walk abreast as this increases erosion at the sides of the path. Sometimes paths are wide enough to walk two abreast but usually it is a case of 'follow-my-leader' and walking in-line ahead. There are many places where paths are twenty or thirty metres wide because people have walked side-by-side or made new paths alongside the original path to avoid the mud where the ground is wet or boggy.

In steep terrain footpaths zig-zag up the hillside so that normal walking and breathing rates can be maintained. While the zig-zags are followed during ascent, there is a frequent tendency to ignore them on descent and travel directly downhill. This is not only bad walking practice but causes one of the worst forms of erosion where the rain drains down the track made by descending walkers and causes serious scouring, taking away the vegetation and topsoil down to the bare rock.

*Popular
paths are
very
seriously
eroded*

Restored footpaths must be used - no walking on either side

Footpaths are precious things which have evolved by a process of natural selection and have stood the test of time, frequently over centuries. Country dwellers, miners, shepherds and drovers balanced the need to keep the distance between two places to a minimum and yet be able to maintain a steady walking rhythm and keep the expenditure of energy to a minimum. The fences and the dry stone walls, which separate the enclosures and fields of the upland areas and the valley floors in wild country, must never be climbed. Stiles and gates must always be used. Footpaths in arable farmland usually make their way around the edges of the fields; short cuts must never be taken through crops and ignorance is never an excuse. Although the Country Code states that gates should be closed after use, there are many farmers who would wish them to be left as found. The need to keep to paths and rights-of-way precludes the use of walking on a compass bearing in normal rural country.

Cycling

The introduction of mountain bikes has created a new interest in cycling which has brought about its own problems of erosion. The heavily treaded tyres cut through the sparse vegetation of mountain and moorland areas, and the steep descents which many bikers make, accompanied by heavy braking, creates deep erosion. Sadly many of the problems arise from bikes being used in the wrong places. Cyclists have no right of access to mountain, moorland or farmland without the express permission of the landowner; neither have they any right of access to footpaths. Cyclists have a legal right to use bridleways providing that they give way to walkers and those on horseback. Extensive use can be made of byways and drove roads, and the Forestry Commission frequently gives permission for use to be made of forest tracks.

Cyclists should confine their cycling to hard tracks and avoid cutting up the vegetation; like walkers, they should avoid taking shortcuts with the consequent erosion

Much of the ill-feeling between walkers and cyclists arises through the silent approach of cycles behind walkers. Cyclists must give walkers audible warning some distance away of their approach; it is not sufficient just to slow down.

Water

The modes of travel associated with water ventures do not present an erosion problem; water does not wear out. Travel is silent and canoes and rowing and sailing boats create little of the wash which causes erosion of the river banks. Canoeing, rowing and sailing on rivers conflict with fishing; as a result, access is severely limited in the British Isles compared with most countries on the Continent.

Litter

Litter is a universal problem in the British Isles and all must be involved in the fight against litter. All litter which arises through camping and travelling must be removed. Litter must not be buried, hidden or burnt but carried away by the participants until it can be disposed of in a bin or receptacle designed for rubbish. It is necessary to carry a few plastic bags for the rubbish from each camp site and day's journey. Experience will tell you how large these need to be. Plastic bags of all shapes and sizes are one of the greatest pollutants of the countryside, fluttering from fences, trees and bushes. They blow away in a moment so keep a tight hold of them at all times. The waste problem may be eased by removing any unnecessary wrappings or packaging from food and equipment before departure. It is usually possible to persuade Supervisors and Assessors to take away the rubbish bags on their daily visits.

There are three stages in the development of an attitude towards litter:

- **The first** when you become aware of your own litter and do something about it.
- **The second** when you become aware of other people's litter.
- **The third** when you are aware of other people's litter and do something about it.

A rubbish bag will help Award participants to perform a community service by removing some, or all, of other people's litter from camp sites, paths and other well-frequented areas.

The same care must be exercised for rivers, estuaries and coastal waters. Litter must never be jettisoned nor should any action taken which will give rise to pollution.

Finally we come to the litter problem which creates the greatest number of complaints to Award Headquarters! Many Supervisors use a letterbox,

Dead-letter droppings

or 'dead-letterbox', system for checking a group's progress across the countryside. The group leaves a message in a plastic bag, or other container, at some significant landmark or where the route crosses a road or track. It usually states the time they reached the place and it confirms, of course, that they have reached that point on their itinerary. It is not the system which causes the problems, but the abuse of the system. Frequently the messages are never collected by the Supervisor. It is unlikely that other countryside users will take offence if the messages were only going to be visible for a few hours but this is rarely the case. If this method of supervising or checking is used, though many would disapprove of its use at all, then **all messages must be collected by the Supervisor, without fail, at the earliest opportunity.**

Fires

Open fires have no place in lightweight or backpackers' camping as indicated in the Chapter 3.2 - Camp Craft, but stoves are frequently used during practice journeys for a 'brew up'. Care must always be exercised in handling stoves and fuel, especially in hot, dry weather when vegetation can be as dry as tinder. The majority of fires are probably caused by dropped cigarette ends, but they can also be caused by a moment's carelessness or even a discarded glass bottle.

Supervisors and Assessors must not allow their vehicles to add to the damage

Joe Cornish

Noise

People go to the countryside to seek peace and quiet and get away from the hubbub of the towns and cities. A blaring radio has no place in the countryside whether at a camp site or in a motor vehicle. Radios in motor vehicles are an ever-increasing pollutant in the countryside. Noise is a pollutant like litter. Country people in general, and farmers in particular, tend to go to bed much earlier than city dwellers; there are many farmers who, as they need to get up at the crack of dawn, will no longer take campers because of the noise they have made at night.

Motor Vehicles

Vehicles are an ever-increasing problem in the countryside and there are certain areas where they are denied access. Great care must be taken to ensure that they are parked correctly where they will not cause obstruction to farmers and country dwellers. Vehicles should not be parked in gateways or where there is access to fields or land, or in narrow lanes where they will cause an obstruction. Supervisors and Assessors usually endeavour to get as near as possible to the participants by vehicle and this in itself may pose parking problems and create obstruction in narrow lanes and tracks.

The increasing use of four-wheel drive vehicles also creates additional erosion. Vehicles frequently have right of access to unsurfaced byways, green lanes and old drove roads, but the unsurfaced roads are churned up and rutted by these vehicles to such an extent that they are frequently impassable to those on foot or horse. The main perpetrators of this erosion are the four-wheel drive enthusiasts, but Supervisors and Assessors must be careful not to add to this problem.

Drivers must remember that it is an offence to drive a vehicle on a bridleway, a private lane, drive or road or on to private or common land without the express permission of the landowner.

Observing, Recording and Presentations

The *Award Handbook* states that all ventures must have a clearly defined and pre-conceived purpose which provides a focus for the venture, and a reflective report related to the purpose of the venture must be produced. The choice of purpose is as wide as the enterprise and imagination of the participants and their ability. The title of this chapter, Observing, Recording and Presentations, sets out the chronological sequence of the process which takes place during and after the qualifying venture. **When a venture is being planned it necessary to consider the presentation first so that the nature of the nature of the observations and the form of recording are suited to the presentation and facilitate its production. Presentations are related to the purpose of the venture.**

Summarising the experiences of the venture and sharing them with others is an important and enjoyable part of the overall experience. When Gold Award holders talk about their Award, ten, twenty or even thirty years after the event, it is surprising how memorable and significant the journey of the Expeditions Section was in their experience.

PRESENTATIONS

Presentations are a vital part of all ventures and it is essential that they should receive attention at the very earliest stage of planning and preparation for the venture. This will enable the presentation to be integrated with the purpose of the venture and all future planning,

Joe Cornish

Observation is enhanced by concentration

preparation and training may help its production and enable suitable forms of observation and recording to be selected.

The Award's *Exploration Resource Pack* is full of information concerning the purpose of ventures, whether they be an Exploration or the traditional Expedition, as well as giving advice on observation and recording. All Units would benefit from having a copy which should be available to the participants for reference. Reference should also be made to Chapter 2.3 - Explorations.

It is the responsibility of the participants to decide on the form and nature of their presentation. It may be written, oral, photographic, audio or video, tape/slide, drama or any other acceptable form or combination of forms. Presentations may be prepared by individuals or the group. Some may decide to prepare individual presentations, while the rest of the group may join together in their preparation. Whatever method is chosen, all presentations should reflect genuine effort from each individual member of the group.

Participants also have the responsibility of deciding who should receive their presentation. Any adult who has been actively involved in the venture, or in its preparation, such as the Assessor, Supervisor, Instructor, Unit Leader or Mentor if the venture is an Exploration, may receive the presentation. The person to whom the presentation is submitted must be sufficiently familiar with all the participants to be able to assess their submissions or contribution in relation to their age, aptitude and ability, but must not be related to any of the participants in the group. This choice enables participants to be imaginative in their presentations and provides greater scope for different methods of presentation. The presentation offers parents, colleagues and others an opportunity to share in the experiences and recognise the endeavours of the participants.

Failure to submit a presentation is the most common reason for the Expeditions Section not being completed. A busy Wild Country Panel, such as that in the Lake District, may frequently find that five hundred presentations or more have not been submitted a couple of months after the end of the Expedition season. Statistics do not reveal how many presentations never arrive, but it is a very considerable number. This is very regrettable as it represents and enormous amount of wasted human endeavour in terms of physical effort, time and money on the part of the participants, as well at the time and commitment of all the adults involved. The almost unlimited choice of reporting styles is designed to accommodate all ranges of aptitudes and abilities from the academic to those with learning difficulties.

All decisions concerning the purpose of the venture and the presentation for Gold ventures must be made before the submission of the green *Expedition Notification Form* to the Wild Country Panel Secretary who will issue a Notification Reference Number for the venture. Notification must be made **at least 6 weeks before the start of the venture** if an Assessor is required and at least 4 weeks before it is for notification only. For ventures in continental Europe and elsewhere abroad, the 'blue' *Expedition Notification Form for Ventures Abroad* must be sent to the Regional Officer via the Operating Authority. The nature of the presentation and the name of the person who will receive it must be made clear on the *Notification Form*. Where participants are going to produce an oral presentation for the Assessor at the end of the venture, the *Notification Form* must make this clear, and details of the arrangements must be agreed with the Assessor.

Environment, Purpose and Presentations

The environment in which the Expedition or Exploration takes place must suit the purpose of the venture. Ventures at Bronze, Silver and Gold levels in normal rural, open or wild country should be remote from habitation. This prevents undertaking any purpose which is based on the use of towns, villages, hamlets or the local population. It is possible, however, to visit a town or village prior to the actual venture. For example, if a group was seeking to follow one of Wordsworth's expeditions in the fells of the Lake District, then a visit to Dove Cottage could take place during the acclimatisation period prior to setting off on the venture.

Presentations

Written presentations, or other forms of presentation submitted after a period of time, have considerable advantages. They provide time for consideration, reflection and a more structured approach. They lend themselves to studies and investigations of an involved nature where a considerable amount of recording has taken place. They can be supported by photographs, drawings, sketches and diagrams. The written presentation also has the great advantage of providing a permanent record of a major achievement.

Other forms of presentation offer scope for imagination and ingenuity. A portfolio of photographs, drawings, paintings or an anthology of poetry, or any form of aesthetic appreciation or expression, all make exciting alternatives. The ever-increasing miniaturisation in electronics offers opportunities for both sound and video recordings and video presentations are becoming increasingly popular and, as with written presentations, they provide a permanent record.

A written presentation, illustrated with carefully chosen photographs, on which effort and care in presentation has been lavished, provides an account which will be treasured in later life, long after the foul weather, the exhaustion and the blisters have been forgotten.

Oral Presentations

Participants wishing to present an oral presentation will need to carry out the preparation for the presentation in addition to the usual observations and recordings.

Oral presentations may be presented to the Assessor at the end of the venture before departure for home, providing that arrangements have been made beforehand. These arrangements need to be completed by the time the *Expedition Notification Form* is sent to the Wild Country Panel Secretary at least six weeks before the start of the venture. The oral presentation related to the purpose of the venture must be separate from the Debriefing which the Assessor will carry out immediately at the end of the venture. The Debriefing is concerned with the participants' immediate reactions to the venture.

The advantages of the oral presentation are based on the event being fresh and vivid in everyone's mind; the sense of accomplishment and enthusiasm is usually plain for all to see. There is the feeling of finishing on a 'high' and there is a completeness about the total experience of the venture. The Assessor can congratulate the group on their achievement and sign the participants' *Record Books*, having had an opportunity to share in their success along with the Supervisor. The problems which arise from participants' being unable to get together again after the completion of their venture are avoided, along with the lack of motivation due to the intervention of holidays etc. which many participants experience in preparing a written presentation weeks after the completion of the venture.

Oral presentations at the end of the venture virtually restrict the participants' presentation to a purely oral one, even through it may be supported by notes, sketches, observations or recordings. There is little time for participants to reflect on or digest the experience, though it is hoped this will inevitably take place later.

An oral presentation requires a considerable amount of time even though much of the preparation will have taken place during the venture itself. Ample time must be set aside and provision made for the presentation. Some form of shelter is usually essential; it could be a frame tent or a minibus. The group must have time to sort itself out and consult with their notes and diaries before making the presentation. The process is usually assisted by a pot of tea or coffee, and helpful questioning by a sympathetic Assessor puts everyone at ease. Oral presentations cannot take place in an atmosphere where the driver, faced with a two hundred mile drive, is impatiently revving-up the minibus engine.

Group and Individual Presentations

The presentation may be prepared either on an individual or group basis. Whichever method is chosen, it must reflect the contribution and involvement of each participant during the venture and afterwards in the preparation of the presentation. Where the presentation is a joint effort by all the group, the person who receives it will need to know who is responsible for the various aspects of the work, such as the drawings, maps and diagrams, any photography, the writing, typing or word processing. Where joint presentations are prepared, each participant should make and keep a copy as it will probably be valued in future years as a record of a memorable achievement.

Groups intending to submit a joint presentation should think carefully of the implications. After the venture has been completed, participants may be going their separate ways on holiday, to work, to university or to college and all efforts to get together may be thwarted. This may account for the failure of many groups to submit presentations and why many participants prefer to prepare individual presentations or arrange to

Paul Smith (Halina/Fuji Bursary)

submit an oral account on completion of their journey. Presentations should be completed as soon as possible after the event while the memories and feelings are still fresh and vivid; preparing a presentation months after the journey can become a very tedious chore.

OBSERVING

The ability to observe is a skill which, like all others, improves with practice for many people look without seeing. Observation is enhanced by concentration and sufficient time to isolate oneself from other matters of concern. Solitude can also be an important factor, and it is possible to detach oneself, even during an Award venture. Some people are able to develop sufficient concentration to observe detail even amidst a group of people. It is also vital to remember that we observe through all our senses: touch, feeling, hearing and smell. It is vital, especially in the world of nature, that we do not limit our sensory input to sight alone as so many frequently do. Experience is always enhanced when it is examined and expressed. As you lie in your sleeping bag drinking your cup of instant chocolate at the end of the day, meditate on the day's happenings; forget about tomorrow, it will take care of itself.

RECORDING

Presentations are always easier to prepare if thought and preparation has taken place before and during the venture. All participants should carry notebooks in which they can keep a log, diary or journal of each day's events. English literature has been enriched by the journals which explorers have kept as a record of their ventures on land and sea, and such accounts are in the great tradition of British exploration. It is important that observations and recordings are concerned with feelings and attitudes rather than with the mundane details of the food consumed and the timings of the journey. What participants think about their venture, the amusing incidents, the anxious moments, the excitement are far more important than what they had for lunch. The disappointments, the feeling of accomplishment, the insight gained about oneself and one's companions are far more valuable than copying large portions out of a tourist guide or reference book.

Participants will need to choose which medium, or media, they are going to use for recording. They may wish to use a method with which they are already familiar or extend their experience by using something new. The Skills Section of the Award Scheme will provide inspiration and competence in the same way that the Physical Recreation Section can

support the technical skills related to the mode of travel. This will not only improve the quality of the presentation but will also increase the satisfaction which comes from the acquisition of another skill and a job well done.

The modern camera is an excellent recording tool. Compact cameras are light, reliable and frequently fitted with automatic exposure and zoom lense. Cameras are a 'must' for most groups, but it is not necessary for everyone in the group to carry one; two cameras are normally sufficient in the hands of competent users, with the rest sharing in the cost of the films and purchasing duplicate prints or slides. This reduces both weight and cost. There will always be those who wish to carry their single lens reflex, but a telephoto lens as big as a milk bottle will make a mockery of the attempts to save weight in the camping gear. Do not learn to use a new camera on a qualifying venture. Photographers should be familiar with their equipment so that they can concentrate on the picture rather than the technology. Use a variety of shots, wide scene-setters, group shots and carefully framed and focused close-ups.

Sound recording has become a possibility with the miniaturisation of recording equipment. More recent advances have made video recording during a venture a viable alternative for some groups. The weight of some modern camcorders is such that, with the appropriate distribution between members of the group, one can be carried quite easily. Carrying sufficient recording tape presents no problem, but at least three battery charges are needed for each tape and it is essential to carry spare batteries and to have some means of recharging them. One Assessor who makes videos of canoeing in Alaska fitted his kayak with solar panels so that he could recharge the batteries. There is no need to go to this extreme; it is far less costly to rely on the Supervisor or Assessor to recharge the batteries, but several will be needed.

Sketchpads, pencils, paints and crayons are also useful for recording and, before the coming of the camera, no explorer worth his salt would have travelled without them. Before the introduction of the camera as a recording instrument, all the larger geographical explorations included artists to paint, draw and sketch the subjects of their attentions. These artists, using the simplest of equipment, have left Britain with a rich heritage of engravings and prints derived from their artistic recording during the 18th and 19th centuries. Recording equipment can be very simple, of negligible weight and no more expensive than a soft lead pencil and a small drawing pad.

Equipment for studies and investigations does not have to be elaborate or expensive and it can frequently be improvised or home-made. Advice is available in the *Exploration Resource Pack* and further details of equipment are included in Chapter 2.3 - Explorations.

One of the most important reasons for deciding on the purpose of the venture at the outset of preparation and training is to enable suitable forms of observation and recording to be selected. These skills can be developed, along with the other skills, during the training and tested during the practice journeys. This will enable any necessary alterations to be made to the method of recording, the form of the subsequent presentation, or even in the purpose of the venture.

EXPEDITION GUIDE

Part 4

ADVICE
AND SKILLS
ASSOCIATED
WITH THE
MODE OF
TRAVEL

Nick Meers / National Trust

Part 4

Advice and Skills Associated with the Mode of Travel

Land Based Ventures

Water Based Ventures

Ventures on Foot

Walking is the preferred mode of travel for the majority of participants within the Award Scheme and will always merit attention as there is something inherently satisfying about completing a challenging journey under your own steam. There is no expensive additional equipment to purchase, other than the clothes you stand up in, and nothing to break down or require servicing other than yourself. There are large stocks of camping gear around the country and the more expensive items of equipment, such as the tent, stove or rucksack, may be borrowed from friends, an Operating Authority or a Youth Organisation, who will need to know that it will be well cared for and returned in a clean and sound condition.

Ventures on foot are usually very successful and predictable despite the vagaries of the British weather. The best way to ensure success is by carefully preparing a venture which conforms to the expectations of the Expeditions Section.

REQUIREMENTS

The requirements for ventures on foot are the same as for all other ventures. See chapter 2.1 - Introduction and Requirements.

these precautions can participants be sure that their equipment will be protected from driving rain and they will enjoy a good night's rest in dry clothing and sleeping bag.

Packing the Rucksack

The distribution of equipment inside the pack is of great importance in reducing discomfort and fatigue. The nearer the load is to a person's centre of gravity the less strain and fatigue it will impose on the body. The design of modern packs helps to do this, but the process must be assisted by heavy items being placed as close to the body as possible and high up towards the shoulders. Packing is largely a matter of common sense and balancing conflicting needs. Since the 'last in, first out' rule applies, items which will be needed during the journey should be placed at the top of the pack or in the side pockets. This will include waterproof over-clothing, spare sweater, gloves, head wear and food which is to be consumed en route.

The sleeping bag and clothing which will not be needed during the walk, although bulky are comparatively light and should go to the bottom of the pack. Participants usually sleep two or three to a tent and the weight of the tent, food, stove and fuel should be divided equally between the occupants. These items should go in next, with the heavy items near to the body and the rest of the gear placed on top.

Fastidious campers put all their gear into their pack and never travel around with their pack decorated on the outside like a Christmas tree. The only exception is the foam sleeping mat which is usually securely attached to the outside and wet clothing which is being dried out.

TRAINING

The Common Training Syllabus as set out in the *Award Handbook* for Bronze, Silver or Gold levels should be used without modification.

Walking

The following advice is directed towards participants who venture into wild country, but it is of equal importance at all levels of the Award as the Expeditions Section is designed to be progressive so that experience is built on the well-laid foundations of the lower levels.

Walking is such a natural process that the majority of participants give little or no thought to the activity, but in this age of mechanised transport, where walking for more than a few hundred metres is a rarity for the vast majority of the population, it merits special attention, especially if the walking involves carrying a pack weighing a quarter of the bodyweight. Participants need to adopt a more purposeful attitude to their chosen mode of travel if their venture is to be successful and not degenerate into just a weary slog. Mental attitude, self-discipline, teamwork and 'getting the act together' are as important as carrying out the physical training required to ensure the necessary levels of fitness.

Participants should agree on a time of departure and everyone should be ready to depart at this agreed time. The group should be suitably dressed for the prevailing weather conditions and rucksacks properly packed. If the departure is from a camp site, the farmer must be paid and thanked, and a last look round the camp site made to ensure that there has not been so much as a matchstick left behind. The first leg of the route must be memorised and maps ready to hand. After making a note of the time on the route card, the group should set out at a slow, steady, deliberate pace. A steady rhythm should be established and all should feel and be conscious of this rhythm. Progress along the route should be checked by noting the landmarks against the map, but this must be done while on

Joe Cornish

the move. Stopping and starting destroys the all important rhythm of the walk, slows down progress and is very frustrating for individuals within the group.

After travelling for about 15-20 minutes, it may be necessary to stop to remove some clothing, especially if it was cold around camp before departure. It is vital not to soak clothing with perspiration. The stop should only take a minute or so and afterwards there should be no more stops until the first leg of the journey has been completed or the group has walked for about an hour. At the end of the leg, or after an hour's travelling, you can reward yourself with a ten minute break. Note the time and then attend to all needs during this break, such as the adjustment of clothing and packs or protection from the sun. The route card should be checked and the map studied to memorise the landmarks on the next leg of the route before noting the time and resuming your journey.

Should the path become steep, the stride should be shortened to maintain the same steady rhythm. If the route should become even steeper, it may be necessary to zig-zag to reduce the steepness of the climb so that the same steady rhythm can be maintained and the heel of the foot can be placed on the ground. It is very tiring walking on the front of the foot; it places great strain on the calf muscles and reduces the grip of the boot which may lead to a fall, especially on steep wet grass.

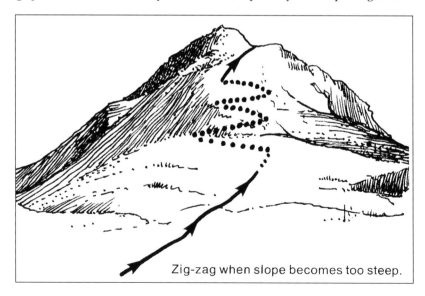

Zig-zag when slope becomes too steep.

Walking down steep terrain may not be exhausting but it is uncomfortable and more slips and falls occur while descending than when moving uphill. When walking on the flat we automatically lean forward to maintain our balance. When walking downhill there is often a tendency to lean back which can have upsetting results; feet shooting from under us and depositing us on our backsides. The weight should be kept well forward, with the knees bent to avoid jarring the joints and it is probably of greater importance to zig-zag downhill than uphill, especially as the loaded pack may pitch the wearer downhill headfirst. **Participants should never run downhill and should be careful never to dislodge stones which may roll and injure someone below.**

If it should start to rain the whole group should stop and put on waterproofs together as it is just as important to prevent clothing being soaked with rain as with perspiration.

All breaks should take place at the end of a leg, at some significant landmark, waymark, checkpoint or at the top of a climb wherever possible. If the group should fall behind time, it may be necessary to reduce the ten minute breaks or even the meal break to catch up with the schedule. It is a good idea to build a little recovery time into the route card to cope with unforeseen circumstances such as bad weather or blisters. If the recovery time is not needed, the time can always be enjoyed amid the scenery of wild country. If continual adjustments are having to be made to the timing schedule it is an indication of a lack of experience in walking as a group or that the practice journeys have not been used effectively.

The time required to deal with the purpose of the venture should be built into the schedule of the route card.

PLANNING AND PREPARATION

Forty years experience and the involvement of several million young people indicate that Bronze, Silver and Gold ventures which take place in normal rural, open or wild country involve few serious incidents, and those which do occur, arise from the actions of the group rather than from the environment or the impact of the weather.

It is an expectation that ventures at all levels of the Award are part of a progression. This progression is concerned with training, experience and competence. Where there is direct entry by a participant at a particular level of the Expeditions Section, care must be taken to ensure that all the training and experience which would have resulted from involvement at lower stages of the Award are fulfilled.

There are three recurring factors, lack of physical fitness, overweight packs and getting lost or going astray due to inadequate navigation, which cause groups to be considerably overdue at the end of the day. In Wild Country Areas this leads to anxiety and concern on the part of Supervisors and Assessors and, on occasions, to the unnecessary calling out of a mountain rescue team.

The chapters on Navigation, Fitness for Expeditions and Camp Craft should help to address these problems. They give advice on the planning of ventures, based on an analysis of the problems which have arisen amongst groups over the years. If this advice is followed, many of the incidents occuring in wild country which result in calling out mountain rescue teams can be avoided, and the probability of success for all ventures on foot will be greatly enhanced. The advice has to be implemented early in the planning process to ensure success.

THE KEY TO PLANNING SUCCESSFUL VENTURES

- Train for physical fitness.
- Divide the journey evenly between the number of days available.
- Start early in the day.
- Make any ascents early in the day.
- Plan alternative routes for use in foul weather.
- If the venture is at Gold level, **travel through, rather than over wild country.**
- Keep the weight of the rucksack to a minimum.

Train for Physical Fitness

Participants need to prepare for ventures on foot by increasing the level of their physical fitness. Only participants who play team games, competitive sport or who carry out fitness training on a regular basis are going to enjoy the challenge of their ventures. The best training for walking is walking; but it must be hard walking, preferably with an increasingly heavy pack. Chapter 3.13 - Fitness for Expeditions provides advice on how this fitness may be attained.

Divide the Journey Evenly Between the Number of Days Available

For a variety of reasons such as the remoteness of the area from the group's home town and difficulties in getting off work, there is a tendency to reduce the distance travelled on the first day, and limit it on the last day so that home can be reached as early as possible. This results in too much mileage being crammed into the middle day or days, unreasonable physical demands being made on the participants and, occasionally, unacceptable levels of risk. One of the reasons for the acclimatisation period before ventures in wild county is to enable ventures to start early on the first day.

Excessive mileage on a particular day usually leads to fatigue and exhaustion, which may lead to hypothermia in the cold wet climate of the British Isles or, on those rare occasions when the weather is very hot, exercise-induced heat exhaustion. Feet receive an extra pounding, groups arrive at the camp site too tired to prepare a proper meal or to make an early start on the following morning.

Start Early in the Day

Journeys should always start early in the morning as this will reduce the chance of being overtaken by darkness at the end of the day. This is particularly important at the beginning and end of the Expedition season when the hours of daylight are limited. It is also a great morale-booster. Groups which start late spend the rest of the day catching up. If a group gets lost there is time to get back on course if they have started off early. If hit by foul weather, there is time to fall back on the alternative route and still reach the destination at a reasonable hour. If all fails and the situation reaches 'panic stations', at least you will be able to panic in daylight which is better than panicking in the dark. After you have had your panic, you will have time to recover your composure, and sort yourselves out.

The above factors apply to all ventures at all levels of the Award. The following have a special significance for those who venture in wild country.

Make Major Ascents Early in the Day

There is great merit in planning a journey so that any prolonged period of uphill walking takes place early in the day when participants are still fresh and, if the weather is hot, the worst heat of the day will be avoided.

Joe Cornish

Participants will be less tired and will have the opportunity to linger over their lunch in high places and to enjoy the panorama spread out below them, knowing that the rest of their journey will be downhill. Should the weather deteriorate they will be heading in the right direction with every downhill step helping to reduce the impact of the weather. Problems tend to arise towards the end of the day so, if the hard work has been completed in the early part of the day, groups are more likely to find themselves in lower, more sheltered terrain and closer to assistance if it is required.

Plan Alternative Routes for Foul Weather

Bad weather should be regarded as normal in wild country, rather than the exception, and all groups should train to cope with it as a matter of routine. There comes a time, however, when experienced hillwalkers and mountaineers are faced with very bad weather, have to think twice and retire gracefully to lower ground. Walking safely in upland areas while cocooned in cloud and mist is a skill which takes years to acquire. Even though the intention is to walk through wild country rather than over it, one cannot escape the rain and the mist. Some Award groups fail to follow this common sense procedure and press on regardless of the conditions for a variety of reasons: they feel that they are 'letting the side down', not facing up to the challenge, throwing away their last chance of gaining their Award, or wasting the time, effort and money they have invested in the venture.

Groups are less likely to be faced with this situation if they can fall back on **'alternative routes'** which will keep them below the cloud base and away from the full force of the elements, to reach their intended destination, and yet still carry out a day's journey of sufficient length to comply with the distance requirements of the Expeditions Section.

An alternative route is one which enables the group to reach its intended destination and meet the distance requirements and yet avoid the full impact of the weather or having to walk through hill fog or cloud which will seriously add to the navigational problems. It is different from an escape route. See Chapter 3.8 - Planning the Route.

Travel Through Rather Than Over Wild Country

High level traverses of well-known ridges and the ascents of summits, though very stimulating and satisfying, are not compatible with the demands of an Award Expedition or Exploration. A distinction must be made between the mountain walker out for the day with sandwiches, spare clothing and a few items of emergency gear in a day-sack and Award participants committed to a pre-planned journey through wild country over a number of days, carrying all their camping gear and food and having to sustain their journey in all kinds of weather. Places such as Crib Goch, Bristly Ridge, Lord's Rake and Striding Edge are unsuitable for Award groups with their 14kg-18kg (30lb-40lb) high-loaded backpacks.

The *Award Handbook* states that **journeys should be through rather than over wild country.** This does not prevent the natural desire of groups to reach the top of some summit, col or pass, but the amount of climbing, or ascent, each day should be limited to around 500 metres. Ascents of more than 500 metres are usually beyond the capabilities of all but the very fittest and experienced young people when carrying a full rucksack. There is always a tendency to over-estimate fitness and under-estimate the effort required to carry a full pack in hilly country.

Assessors will look upon an excessive amount of climbing each day with suspicion. Distances are mandatory for ventures on foot and the only way the physical effort can be regulated is by controlling the amount of

climbing each day. If a group has a desire to reach the summit of some peak which is over 500 metres above the valley floor and can be reached without travelling along exposed ridges and edges, then it would be wise to restrict the amount of climbing on the following day. By keeping well away from the well known ridges and edges when carrying a loaded rucksack, participants will not only be safer but will avoid the crowds which frequently queue up in these places. **Solitude not altitude** is the by-word within the Expeditions Section.

Keep Pack Weight to a Minimum

The weight of a participant's pack will determine, more than anything else other than physical fitness, how much the venture will be enjoyed. The weight of a pack should not exceed one quarter of one's bodyweight. Once the basic items of emergency and camping equipment have been assembled, extra items will not add to safety, they will just add to the load. An excessive load gives rise to fatigue and exhaustion and slows progress which, in bad weather, increases the probability of hypothermia. Great personal discipline must be exercised by ruthlessly

---

Weigh your rucksack during and after packing. The lighter the better - not more than a quarter of your bodyweight!

eliminating all unnecessary items. There must be consistency - there is little point in counting tea bags, finding the lightest tent and then carrying a giant lump of soap, a bath towel, a year's supply of toothpaste and talcum powder and a single lens reflex camera with a telescopic lens weighing a kilogram. There need never be more than a sliver of soap and a small well-used hand towel. All packs must be weighed before departure and this is most easily done on the bathroom scales while the equipment is being packed. Just as fitness training was one of the first preparations, so keeping the pack weight down is one of the last actions before departure to ensure that the venture will be enjoyable.

The factors listed above should form the basis of route planning but a few other factors have to be taken into account.

Long Distance Footpaths

The use of long distance footpaths for practice or qualifying ventures is not considered to be sufficiently demanding in the planning or navigation of Award ventures. Extensive literature, maps and guides deprive venturers of the opportunity to plan their route and many of these paths are waymarked along their entire length. The sheer volume of walkers using these paths does not fit in with the ethos of the Award Scheme. It causes considerable erosion, has an irreversible effect on the ecology around these paths and serves to destroy the solitude which can contribute so much to the feeling of achievement in wild country travel, and which is such an important ingredient of successful expeditioning. Long distance footpaths should not be included in routes for Award ventures, unless short sections are unavoidable to link up with other paths or rights-of-way.

Use of Private Land

Although ventures often take place in remote areas where, at first sight, there is little evidence of agricultural or other activity, all land has some use to its owner and it is therefore important that prior permission should be sought. At certain times of the year, groups may not be welcome for both agricultural and sporting reasons. It is important to note the dates of such seasons in the Wild Country Areas of the United Kingdom.

- **Lambing** - March, April and May.
- **Grouse Shooting** - 12 August to 10 December.
- **Red Deer Stalking** - 1 July to 15 February.

These interests have economic significance and must be respected. Additionally, for deer stalking, long range high velocity rifles are used which can be dangerous at very great distances. To provide landowners and farmers with credentials and to safeguard the good name of the Award Scheme, a simple *Expedition Identification Card* is included in all *Gold Entrance Packs*.

In certain Wild Country Areas there is a tradition of free access to the upland areas and the problems of access tend to be greater in the valleys and on the lower slopes. The valley floor and sides are enclosed by small fields which provide fodder and winter pasture. The fields are usually surrounded by characteristic stone walls which should never be climbed. It is essential to use the 1:25 000 scale maps wherever possible for route planning and navigation, as the field boundaries are shown on these maps, but not on the 1:50 000 scale *Landranger* maps. They also show the tracks and paths which give access to the more open terrain of higher ground. Rights-of-way are clearly marked and paths are shown in greater detail, making them easier to follow and therefore avoiding problems with the landowners.

Valley Camp Sites

By custom, groups nearly always spend the night in valley camp sites which are not far from a road or track with vehicular access. This considerably eases the burden on Supervisors and frequently hard-pressed Assessors when making early morning and evening visits. More importantly, should participants be taken ill during their venture this usually tends to happen at the end of the day's walking or during the night. The proximity of a telephone in a nearby farmhouse has proved to be of very significant help over the years. Valley camp sites tend to be more sheltered and less exposed to the elements.

Cycling

Cycling ventures have never enjoyed the popularity which might be expected considering the number of young people who have bicycles. There may be a simple explanation such as difficulty in finding a sufficient number of friends with the same kind of bikes within an Award Unit; many may have suitable bikes but few possess the essential panniers. A renewed interest in cycling brought about by the popularity of the mountain bike may change all this and help to bring about an increase in the number of cycling ventures.

Over 70% of all bicycles sold are mountain bikes, though the original mountain bike has undergone so much development and diversification that it is more correct to call them 'all-terrain' bikes or ATBs. The rapid development of the mountain bike into the all-terrain, all-surface, go-anywhere bike, with all purpose tyres, has made it impossible to divide bikes into just two categories - the mountain bike and the touring cycle - as was previously the case within the Expeditions Section. To encourage cycling Expeditions, changes have been made to the conditions.

The mileage requirements have been changed to a minimum number of hours of travelling, bringing this mode of travel into line with sailing, canoeing and horse riding ventures.

Some with touring bikes will prefer to confine their venture to surfaced roads

National Trust

This change will enable participants to pick and choose routes suited to their bikes and their inclinations. Some, with traditional touring bicycles, will prefer to confine their ventures exclusively to surfaced roads, while others may want to use their all-terrain bikes on bridleways and tracks and avoid using surfaced roads except to join lengths of track and bridleway together.

Bikes lend themselves to Explorations just as easily as any other mode of travel and a little thought will reveal purposes for Explorations which are ideally suited to the use of the mountain bike or the touring cycle. There are excellent opportunities on the Continent and an area like the Harz Mountains offers unlimited access on the Wanderwegs, though bikers are no more popular with walkers there than they are in the British Isles.

To avoid the necessity for two separate chapters, one dealing with the traditional touring bicycle and the other with the all-terrain bike, and at the risk of a little confusion, they will be referred to as 'mountain bikes' or 'cycles', or more usually as just bikes, as this appears to be the most widely used terminology.

Though road distances are generally referred to in miles, and this is unlikely to change in the immediate future, both metric and imperial units are used. Participants should use the unit of their choice, but should not mix the two on their tracings or route cards.

The text in this chapter is biased to Expeditions which take place in wild country as these ventures are likely to make the greatest demands on the group, Supervisor and Assessor but the advice applies to cycle ventures at all levels of the Award.

MOUNTAIN BIKES

Award Headquarters receives more enquiries concerning mountain bikes than any other mode of travel as well as receiving more complaints concerning their use. Mountain biking is a 'conflict' sport, like many other sports such as fishing and canoeing, and the use of mountain bikes within the Award Scheme must be carefully regulated. Walkers, in particular, take grave offence at bikers approaching them silently and at speed on narrow paths, endangering life and limb. Great care must always be taken when approaching other people in the countryside, especially when approaching from behind. Audible warning of approach should always be given some distance away and bikers should always slow to walking pace and, if necessary, stop and dismount. The use of bikes within the Award Scheme must be legal and acceptable to other people, and the greatest care and courtesy must always be exercised by participants. Bikes can usually be insured as part of normal household insurance policies at a very small extra premium, which will not only cover the loss of the bike but may provide important third party cover as well.

- Cycling on any footpath is illegal, as it is on pavements.
- Cycling OFF footpaths in open or mountainous country is also prohibited even though de facto access may be available to those on foot.
- Cyclists do have a legal right to use bridleways though they must give way to horse riders and pedestrians.

Assessors will always check routes to ensure that they do not contravene these requirements.

In some areas of the country, considerable use is made of Forestry Commission and Water Authority tracks by agreement with the landowners. This can provide challenging terrain and demanding navigational problems with little or no conflict with other countryside users. The Forestry Commission has produced a Code of Practice which should be observed.

D.W. Elson

REQUIREMENTS

The requirements for cycling ventures are the same as those for all other ventures. See Chapter 2.1 - Introduction and Requirements.

CONDITIONS

BRONZE

Expeditions must involve travelling for a minimum of four hours on each of the two days in addition to other planned activity. Expeditions should take place in normal rural country which is unfamiliar to the participants. Routes may involve minor roads only and should include lanes, bridleways or unsurfaced tracks or any combination of these.

SILVER

Expeditions must involve travelling for a minimum of five hours of each of the three days, in addition to other planned activity. Expeditions should take place in normal rural or open country which is unfamiliar to the participants and is more demanding than that which should be used at Bronze level. Routes may involve minor roads only and should include lanes, bridleways and unsurfaced tracks or any combination of these.

GOLD

Expeditions must involve travelling for a minimum of six hours of each of the four days, in addition to other planned activity. Expeditions must take place in wild country and unpopulated areas. Routes may involve minor roads only and should include lanes, bridleways and unsurfaced tracks or any combination of these. The route and the surrounding area must be unfamiliar to the participants.

CONDITIONS WHICH APPLY TO ALL VENTURES

The same type of bike, mountain or touring, must be used for the practice journeys as for the qualifying venture.

All participants must know the Country Code and the Highway Code, especially those areas concerned with cycling. In addition, groups travelling off surfaced roads must use the Mountain Biking Code. If Forestry Commission land is used, a copy of their code must be obtained and the requirements and advice adhered to.

GENERAL CONSIDERATIONS

Many young people involved in the Award Scheme have mountain bikes but they should not underestimate the challenge presented by cycling ventures, especially if they are going to use their bikes off-road. Because mountain bikes are the more popular, advice is slanted towards their use off the road, but it would be wrong to think that their use on surfaced roads is a soft option - they are very heavy compared to traditional touring cycles and the heavy tyres provide considerable rolling resistance.

Participants must not underestimate the damage which can be caused to their cycles through reckless or thoughtless riding. Cyclists must exercise the same level of care for their bikes as horse riders would for their horses. It is often prudent to dismount to protect the cycle, even though the section could be ridden, to ensure that they are able to complete the venture.

EQUIPMENT

Note: Any participants who are purchasing bikes, panniers, helmets, shoes or other equipment or clothing specially for an Award cycling venture should seek advice from a suitably experienced person to avoid unsatisfactory purchases which may be regretted later.

General Equipment Related to Mountain Bike or Touring Cycle

All cycles must be in a sound and suitable condition and be fitted out for the conditions in which they are to be used.

Food and equipment will have to be carried in panniers. These tend to be expensive, though some Operating Authorities have a stock which may be borrowed or hired. Large capacity panniers which fit on racks over the rear wheel should suffice for an Award venture. Panniers are also available for the handlebars and front wheel. Weight must be kept to the minimum with the same ruthlessness as in foot ventures; weight is always a handicap. You may have to lift the bike and panniers over obstacles on bridleways.

The commercial fastenings which secure panniers to their racks fail sooner or later - usually sooner. Each pannier must always be supported by at least two bungee cords, or cycle inner tubes, wrapped around the frame and pannier in such a way that they will absorb the constant shock loads to which the panniers are subjected.

A repair kit to carry out running repairs should include:
• Puncture outfit, spare valve, spare inner tube, tyre levers and pump.
• Spare brake blocks and brake cables.
• Spare gear cables.
• Spare chain links and chain tool.
• Two or three spare spokes and a nipple tool.
• Spanners, Allen keys, pliers and screw drivers necessary for the above.
• Some rag and a very small plastic container of 'Swarfega' or liquid detergent.

A lock and chain will be needed to prevent the bike being stolen. If a combination lock is not being used the spare key should be kept in a different place, or carried by another member of the group. Many riders remove the front wheel if it has a quick release and chain it with the back wheel, while some remove the seat and take it with them.

Front and rear lights are essential equipment even though there may be no prior intention to ride at night. A headtorch is most effective for seeing where you are going at night in off-road situations and are most useful around the camp site, keeping the hands free. Cyclists generally carry water/fluid bottles.

Clothing

Helmets: Helmets must always be worn on Award ventures whether on or off the road. Heads are the most vulnerable part of the body and most likely to be injured in a fall and, since we have only one, it deserves the best care and protection we can provide. Modern helmets are light and well ventilated. They should fit securely and not interfere with the rider's vision. Brightly coloured ones make the rider more visible, and reputable brands carry the British Standards kite-mark.

Shoes: Specialised biking shoes have hard stiff narrow soles with shoe-clips but lightweight mountain or hiking boots will suffice, as will trainers with substantial soles, especially as there may be a considerable amount of walking involved in the venture. Shoes for cycling must be broken in with the same conscientiousness as walking boots, if blisters are to be avoided.

Gloves: They not only keep hands warm, prevent blisters and provide grip, but protect the hands from cuts and abrasions when they are used instinctively to break a fall in a spill.

Clothing: Brightly coloured clothing should be worn which is highly visible in traffic. Reflective strips, 'Sam Brownes' or patches are essential for dusk and night riding even though there should be no need for this during an Award venture. There is a wide variety of clothing which has been specially developed for the cyclist, but it is possible to manage quite adequately with normal clothing not dissimilar to that used by the hillwalker. The special pants developed for the cyclist out of spandex-lycra usually have padded crotches, a welcome luxury on long gruelling ventures.

Dressing for biking is not dissimilar to dressing for ventures on foot and the same principles apply - layers of clothing which can be removed or added to as temperature and work-rate dictate. Chapter 3.1 - Equipment should be referred to, but it is essential to avoid clothing which is so loose and floppy that it might cause an accident. Traditional materials such as wool or cotton may be used to keep warm or cool as circumstances demand or, alternatively, use the modern synthetics which do not absorb as much water.

Clothing will have to cope with conditions around the camp in the evening and morning and waterproof overclothing should be regarded as essential. If anoraks or cagoules are worn while cycling, it is vital that the hoods do not impair vision or hearing. Track suits are a useful addition to any wardrobe, particularly if wearing cycling shorts.

The list of personal and emergency equipment, and personal and group camping gear described in Chapter 3.1 can be used without any amendments.

TRAINING

The Common Training Syllabus, as set out in the *Award Handbook*, must be used together with training to the standard of the National Cycle Proficiency Scheme, obtainable from RoSPA (see Appendix). Participants should know the Highway Code, especially those sections which are relevant to cyclists, as well as the Country Code. They should also be familiar with the Cyclists' Touring Club publication 'Cycling Off-Road and the Law' by Neil Horton if they intend to leave the public highway. If appropriate, The Right Track Awareness Programme should also be followed.

Whilst Instructors who are able to deliver the Common Training Syllabus and train participants for ventures on foot will be able to deliver most of the training for cycle ventures, **there is a need for specialist instruction in the skills relating to the mode of travel, whether it is for mountain bikes or touring cycles.** Many participants will have considerable skills and expertise with regard to both riding and maintaining bikes, but this competence will still have to be evaluated and a complete programme of training must be provided. There are no qualifications or levels of experience stipulated for those who give instruction in this area of outdoor pursuits, but instruction must be delivered by competent and experienced cyclists. The local Cycling Club may be able to assist in this process.

For mountain bikes specific skills and techniques are required if they are to be used efficiently. These riding techniques are very different from those for touring bikes and it is important that participants have instruction in these techniques and do not have to rely on being self-taught for, in a group, there will always be individuals who have gaps in their competence.

SAFETY

All participants must be competent in handling a fully loaded touring cycle or mountain bike. Great care should be exercised on all roads, especially on narrow country roads and lanes with their blind corners and bends. Riding in a group is a totally different experience from riding on one's own and individuals are more in danger of injury from each other than from other road traffic. Groups should always ride in single file and keep two bike lengths apart, extending this distance to at least twenty metres when descending hills or in difficult terrain.

Cycle ventures involve greater distances and higher speeds of travel than ventures on foot and it is much easier for members of a group to become separated from each other, even discounting the problems of traffic and route finding. Great care must be taken to stay together as a group, especially if members are very tired at the end of the day. It has been suggested that the whistle, which is part of every hillwalker's emergency equipment, is the only practical way of attracting attention by the backmarker who is in danger of being separated from the rest of the party ahead.

Where there is a tendency for a group to spread out, it is helpful if the strongest rider brings up the rear and the slowest moves to the front. This will enable the group to stay together and assist the self-discipline of the strongest riders.

SETTING UP THE BIKE

Whether the bike being used for the venture has been bought, hired or borrowed, it must be suitable for the nature of the venture and the right size for the user. Young people are just as much in need of expert advice or instruction using bikes as in any other form of travel within the Expeditions Section. Size is important for comfort and safety; riders should be able to stand comfortably astride the bike with a gap of one or two inches between crotch and crossbar on touring bikes and nearly double this on mountain bikes.

A bike of the right size and dimensions having been acquired, it is essential that the bike is correctly adjusted to the rider's build. This is particularly important with bikes which have been borrowed or hired. Many young riders suffer endless discomfort and fatigue because they have not received advice from an experienced person in adjusting the saddle and handlebar height to their own particular needs.

MAINTENANCE

A maintenance course must be built into the training programme for groups intending to go on cycle ventures as bikes need to be carefully maintained if they are to be trouble-free and reliable. The more a bike is used off-road the more care and attention it will need. Many Outdoor Activity Centres using mountain bikes ceased to do so when the full implications of maintenance became apparent. All members of the group need to be able to:

- Keep the bike clean and oiled.
- Mend a puncture including removing and replacing the rear wheel.
- Adjust and replace brake blocks and a worn brake cable.
- Adjust the chain tension and repair a broken chain.
- Adjust derailleur gears which have been knocked out of alignment and replace a snapped gear cable.
- Remove and replace a broken spoke.

If bikes are properly maintained and serviced before the start of a venture, there should be little need for other repairs or adjustments en route to such things as hubs, bottom bracket or steering.

FITNESS

The difference between those who are physically fit and those who are not is probably more marked in cycling ventures than in those on foot. Cycling twenty miles a week to work or to school is inadequate preparation for a cycling Expedition. A fitness programme which becomes progressively more strenuous is an essential part of the preparation. The best training for cycling is cycling, and it should always involve a lot of hill climbing and load carrying. Those who are unable to cycle on a regular basis should use a stamina building programme of at least twenty minutes a day, two or three times a week as a substitute.

Riding off-road has its own specialised techniques which must be mastered - even such a fundamental process as braking may be completely different. Those who ride regularly in rough terrain may lessen the chances of serious injury by learning to relax and absorb the shock when they fall. The use of the cross-over gears must be fully understood and the use of derailleur gears must be automatic so that the optimum pedal rate can be maintained at around 80-90 revolutions per minute. These techniques should be practised for off-road ventures, first without a load and then with a full load of equipment. The load will not only limit technique, but restrict journeying to bridleways and unsurfaced tracks and lanes. Concentration is essential at all times when riding on and off the road, not only for the rider's safety but to protect the bike from damage - it would be very frustrating if a venture had to be aborted just because of a moment's lapse of concentration.

In all probability the bikes will have to be transported to the point of departure. This will almost certainly be the case for Gold ventures which must take place in a designated Wild Country Area - there is a wide variety of cycle racks available for cars and minibuses.

Packing

Backpackers tend to place the heavy items high on their shoulders. For cyclists the opposite is the rule - heavy items are placed in the bottom of the panniers so that the centre of gravity is kept as low as possible. The weight must also be carefully balanced between the panniers on either side. Spare clothing, sleeping bags and food must be waterproofed by being enclosed in poly-bags, for it would be wrong to expect panniers to be any more waterproof than rucksacks. Waterproofs, maps and snacks to be eaten en route should always be easily accessible on the top of the equipment.

Equipment must not be carried on the back in a rucksack, as this will not only make the rider unstable with a high centre of gravity, but will increase the probability of going over the handlebars in an emergency stop. A bumbag or a small day-sack is acceptable for items needed en route. Small day-sacks could be used by the group for day training rides where panniers are not fitted to the cycles.

Planning the Route

As a starting point for planning, an average group might expect to cover the following distances on surfaced roads, with a loaded bike, during a cycling Expedition:

Bronze:104km/65 miles Silver:176km/110 miles Gold:240km/150miles

As a rule of thumb 1 kilometre/mile travelled on an unsurfaced bridleway or track is the equivalent effort to 2 kilometres/miles on surfaced roads. Groups will be able to gauge the distances they can travel during practice journeys.

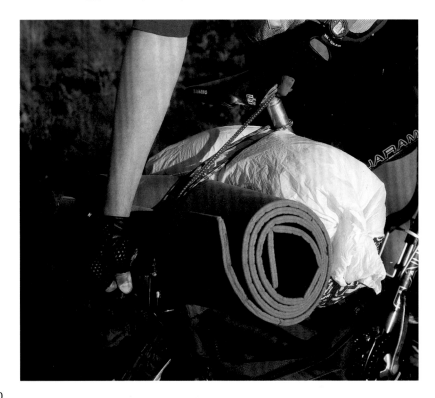

As with ventures on foot, no specific allowance is made for height climbed, but the minimum mandatory travelling time allows for a designated Wild Country Area.

With the considerable distances involved with cycle ventures, an opisometer, or map measurer, is a great help. Choose one with a straight shaft as they are easier to twirl between finger and thumb when following the route. Make sure the needle is set to zero, then trundle it along the route you wish to measure. At some scales it is possible to read the distance directly off the dial, but at others it is necessary to trundle them in the opposite direction along the scale on the edge of the map.

Planning a route at **Bronze level** should be a relatively simple process. It is necessary to find an interesting area of rural countryside which will suit the purpose of the venture, obtain a 1:50 000 map of the area and devise a route of at least the required duration, which avoids major roads, towns and villages with a camp site somewhere near the middle of the journey.

At **Silver level** the task becomes a little more difficult as the minimum hours of travelling make the distances to be covered greater. While rural and open country may be used, the route should become demanding than one used for a Bronze venture. Keeping away from towns and villages makes considerable demands on the participants' planning, but there are over 14,000 miles of bridleways in the country which may be tackled.

Ventures at **Gold level** require careful planning and the preliminary planning is best done on a road map or atlas. It will soon become apparent that some of the Award's Wild Country Panel Areas are too small to accommodate the 6 hours of travelling a day of a Gold cycle venture and this will restrict the choice. Wild Country Areas such as the Cheviots, Cumbria and Mid Wales can accommodate Gold cycle ventures using a circular route which enables the same Assessor to be used for the whole of the venture, benefits all concerned and is reassuring for the participants. It is always possible to move from one Panel area to another where the areas are contiguous such as in Wales, the Pennines and Scotland including the Borders.

Route cards for cycle ventures are prepared in the same way as for ventures on foot but the 'legs' are longer. A typical route card for a walking venture at Gold level might have five or six legs, about four or five kilometres in length for each day, with the corresponding number of checkpoints. Cycling Expeditions involve a greater distance thereby increasing the length of each leg to around ten or twelve kilometres so that the number of legs is restricted to five or six, and there is not a great bundle of paper work.

SUPERVISION

The Supervisor must be familiar with the contents of Chapter 6.2 - Supervising Qualifying Ventures and be approved by the Operating Authority. The supervision of cycling ventures does not differ greatly from ventures on foot except that the distances involved are three or four times greater. This makes considerable demands on the mobility of the Supervisor who may well have to plan to move base during the venture. This would be essential in ventures which have followed a linear route such as using minor roads to keep as close as possible to the Pennine watershed. In the Kielder Forest, in the Cheviots Panel's area, a circular route might be planned making it possible for the Supervisor to remain at the same base. Where the participants are engaged in a venture which is almost entirely off-road, Supervisors may well need to use mountain bikes themselves.

ASSESSMENT

The Assessor must be familiar with the contents of Chapter 6.4 - Assessment, briefed for the role and should be accredited if possible.

The assessment of cycling ventures presents no particular problems to the Assessor other than that of the very considerable distances covered at Silver and Gold levels. Many Wild Country Expedition Panels have very competent cyclists amongst their members who specialise in looking after these Expeditions. Panels which have a tradition of assessing off-road cycle ventures which take place in forested areas have competent off-road cyclists among their members who will reach camp sites in the most inaccessible of places.

Where ventures cover one or more Panel areas it is customary to pass the assessment over to an Assessor in the adjoining areas to save travelling time and expense. Usually the Panel Secretary of the Area where the venture starts will liaise with the Secretaries of the other Wild Country Areas. **Where an Accredited Assessor from an Operating Authority is being used, all Wild Country Panels on the route must be notified in the usual manner.** Where such practices take place, the group, the Supervisor and all concerned must be fully briefed at the **First Meeting** about all the arrangements which have been made.

Horse
Riding

orse riding ventures present a very special challenge to participants within the Award Scheme, especially at Silver and Gold levels. All who achieve success deserve special commendation for the challenges which they have overcome. It may not be wise to undertake a riding venture unless participants are qualified to the standards listed below, or of equivalent experience, **and are involved in riding in rural surroundings on a regular basis.** Some of the problems associated with riding ventures are outlined below so that all participants will have a clear understanding of the challenges which they face when tackling this mode of travel and can train and prepare to ensure their success.

Horses are creatures of habit with their own personalities; they do not always take kindly to different surroundings and are capable of expressing their disapproval in forceful ways. Unless horses are accustomed to the other horses in the group it may give rise to problems when they are being stabled or turned out at the end of the day.

It is frequently difficult and expensive to hire suitable horses. Horses are affected by weather conditions as much as their riders; they may not, for instance like thunderstorms, even though they may be infrequent during ventures. Horses in the first quarter of the twentieth century were used to working in harsher conditions and more demanding situations than horses used solely for recreation in the last quarter of the century. Horses and ponies need to be prepared and conditioned for Award

Expeditions as thoroughly as their riders. They must be fit and accustomed to the demands which will be made upon them during a journey which will involve four, five and six hours' travelling each day at the Bronze, Silver and Gold levels of the Award. At Bronze level one's own horse or pony may be suitable but at Silver level, and especially at Gold level where ventures take place in wild country, horses must be very fit, suitable for and used to meeting the demands made by the terrain involved in the journey.

REQUIREMENTS

The requirements for riding ventures are identical with those for all other ventures as given in Chapter 2.1 - Introduction and Requirements. Participants are expected to be as dependent on their own resources as much as possible, but it is acknowledged that it may be necessary to pre-position camping equipment, food, fodder and the ancillary equipment associated with the horse. Although the country must be unfamiliar to the participants, a limited amount of reconnoitering may take place to ensure that tracks and bridleways can be negotiated on horseback. The paragraph on 'general considerations' (page 308) provides details of the limitations governing these exceptional requirements.

CONDITIONS
BRONZE

Duration: Expeditions must involve travelling for a minimum of 4 hours on each of the two days in addition to other planned activity.

Environment: Expeditions should take place in normal rural country which is unfamiliar to the participants. Routes should involve lanes, tracks and bridleways. Villages should be avoided where possible.

Competence: Training to Pony Club 'C' Standard. All must be competent in ensuring the well-being of the horse for the duration of the venture, be able to recognise difficult and dangerous going and know the action to be taken in the event of an accident to horse or rider.

SILVER

Duration: Expeditions must involve travelling for a minimum of 5 hours on each of the three days in addition to other planned activity.

Environment:	Expeditions and Explorations should take place in normal rural or open country which makes more demands on the participants than that which would be used at Bronze level. Routes should involve lanes, tracks and bridleways. Every effort should be made to avoid villages. Wild country may be used but the conditions for Gold ventures must be applied.
Competence:	Training to Pony Club 'C' standard. All must be competent in ensuring the well-being of the horse for the duration of the venture, be able to recognise difficult and dangerous going and know the action to be taken in the event of an accident to horse or rider.

GOLD

Duration:	Expeditions must involve travelling for a minimum of 6 hours on each of the four days in addition to other planned activity.
Environment:	Expeditions or Explorations must take place in wild or open country. Routes should involve lanes, tracks and bridleways. Every effort must be made to avoid villages.
Competence:	Training to the equivalent of Pony Club 'C+' Standard. All must be competent in ensuring the well-being of the horse for the duration of the venture and be able to recognise difficult and dangerous going and know the action to be taken in the event of an accident to horse or rider.

CONDITIONS WHICH APPLY TO ALL VENTURES

A knowledgeable person should ensure that the horses are fit before the group sets out and after the venture has ended.

At the beginning and end of each day participants should ensure that their horses are sound and fit enough to continue with the venture.

Camping equipment, food, fodder and items associated with the needs and care of the horses, such as grooming tools and buckets, may be pre-positioned for riding ventures, but the emergency equipment listed in Chapter 3.1 and the participants' own sleeping bags and a day's food must always be carried.

For all qualifying ventures the distance requirement starts at a pre-determined point in normal rural, open or wild country and may not include the distance travelled from home or through urban areas.

All participants must know the Country Code and the Highway Code, especially those elements concerned with riding. In addition, they should have knowledge of the Riding and Road Safety Test published by The British Horse Society.

GENERAL CONSIDERATIONS

All groups should endeavour to be as independent of outside help as possible. In the past, participants have constructed very effective lightweight saddlebags and bed-rolls at little cost. In their desire to make themselves self-reliant, they have not only increased the challenge and acquired new skills, but have also increased the satisfaction which they have derived from their venture. Such efforts must always be commended.

It will probably be necessary to pre-position most of the food, fodder and equipment needed for a venture because it is not possible to carry it all on the horses. The only way that it would be possible to carry the equipment would be by means of a pack-horse and it is unlikely that one would be available for those engaged in Award ventures. Leading a pack-horse from another horse is a technique in itself which can only be acquired through considerable practice.

The equipment which may be pre-positioned must conform strictly to the usual lightweight camping gear, suitable for carrying in a rucksack, and used in all Award ventures. Frame tents and sophisticated cooking equipment must not be used. Camping and cooking must take place in the usual two or three person units. Bivouacking in a hay loft or barn is permitted, but extreme care must be exercised while cooking. All participants must be totally self-sufficient and free of any adult intervention.

The pre-positioning of equipment will require the involvement of a Supervisor but this must not extend beyond the delivery of the equipment and the usual duties of supervision.

It is essential that each participant carries all the emergency equipment listed in Chapter 3.1 - Equipment, in addition to their sleeping bag, food for the day and emergency rations to ensure that their safety will not be compromised if an emergency prevents them from reaching their destination. This equipment, along with specific items concerned with the care of the horse should be carried in a saddlebag or lightweight day-sack. Many riders find day-sacks uncomfortable because they rub and cause sweating. If one is used it must be small, so that it does not bump on the back of the saddle or impede the rider, and should be fitted with a waistband to stop it bouncing around. The ones designed for runners which have elastic compression straps are especially good. Alternatively a bumbag, which does not interfere with the rider at the walk and trot, is useful for bits and pieces needed during the day.

The participants will need to reconnoitre the chosen route before embarking on their venture. The depiction of a bridleway or track on an Ordnance Survey map does not necessarily mean that it is possible to negotiate the route on a horse. Many bridleways are obstructed, possibly as many as a third of them - it is much harder to lift a horse than a mountain bike over an obstruction! Short of carrying a pair of bolt cutters and a chain saw in the saddlebag there is no solution other than finding a different route and reporting the obstruction to the relevant local authority.

Using a map while in the saddle can be difficult, especially if the map needs to be spread out; horses may be startled if they see maps flapping about in their peripheral vision. In addition to the usual route card, an itinerary should be prepared which can be used to follow the route without the need to dismount to read the map.

EQUIPMENT
General Equipment Related to the Horse
The equipment related to the needs of the horse falls into two categories:
• Items which may be essential for the well-being of the horse and should be carried during the day.
• Items which may be pre-positioned.

Items to be carried during the day:
• Head collar and rope.
• Hoof pick.
• First aid equipment for the horse which should include antiseptic cream or powder, cotton wool and a leg bandage.
• Fly repellent.
• Bailer twine to tie up gates, repair bridle etc. A penknife is standard personal equipment.

Items which may be pre-positioned:
• Fodder.
• Grooming tools.
• Water and feeding buckets.

Clothing
Every participant in an Award riding venture must wear an approved protective riding hat with safety strap which must be well fitting and securely attached so that it will give protection in the event of a fall. Gloves are recommended. Dress for trekking may be quite informal but it is essential that clothing and footwear must be suitable for the activity and conform to current good practice to ensure the participants' safety and well-being for the duration of their journey.

The usual outdoor activity policy of using layers of clothing which may be removed or added to as weather conditions and body-warmth dictate, should be followed; trekking is no exception. Waterproof outer clothing such as a cagoule may be worn, but should not impede vision, hearing or control of the horse, and should not flap in the wind and possibly startle the horse.

Personal and Emergency Equipment

The personal and emergency equipment listed in Chapter 3.1 can be used without modification for riding ventures. If the group is able to carry a small lightweight tent into which they could all crowd in an emergency, then one should be carried with the weight being distributed amongst the group. If not, the bivvy bag (large poly-bag) and the spare clothing assume even greater importance. There are lightweight waterproof covers available on the market under which a whole group can huddle together, drawing the lower part underneath themselves to give additional protection. This equipment should be carried in the saddlebag or a light day-sack.

Personal Camping Equipment

This equipment may be pre-positioned with the exception of the sleeping bag which must be carried.

Group Camping Gear

This equipment may also be pre-positioned.

TRAINING

The Common Training Syllabus as set out in the *Award Handbook* must be used. Training related to riding ventures for each level of Award should be to the appropriate standards listed below:

Bronze and Silver	**Gold**
Pony Club 'C' Standard	Pony Club 'C+' Standard
Riding Club Grade 2	Riding Club Grade 3
British Horse Society Progressive Test 10	British Horse Society Progressive Test 12

It is not essential for participants to have achieved these standards as there may be other equivalents, but these levels of competence should ensure the safety of the riders and the care and well-being of the horses, as well as equipping the participants with the necessary skills to carry out their journey with confidence. Instruction should be delivered by a person who is appropriately qualified, or of equivalent experience and approved by the Operating Authority. Instruction must be included in those aspects of riding which are specifically involved in pony trekking and the care of the animals used in the activity, in addition to the competencies listed in the tests above.

Planning is of great importance in riding ventures as it is not sufficient to find a well-disposed landowner who will permit the participants to camp on his property. The landowner must be prepared to provide water and secure pasture or stabling for the horses, in addition to a field in which the participants may camp. Camp sites on farms which can provide the necessary facilities must be separated by the mandatory number of riding hours and be linked by means of open country and/or bridleways, lanes and tracks which are not obstructed. **This will require a considerable amount of prior research on the part of the members of the group.**

SUPERVISION

The role of the Supervisor takes on an added significance in riding ventures as they not only have the responsibility for the safety of the participants but also for the welfare of the horses and for pre-positioning some of the equipment. Supervisors must be familiar with and able to fulfil their responsibilities to the participants as set out in Chapter 6.2 - Supervising Qualifying Ventures and be approved by the Operating Authority. In addition, Supervisors must be sufficiently well qualified and experienced to ensure that the horses are properly cared for.

The role of the Supervisor in riding ventures may be very similar to that of those who supervise mountain bike or canoeing ventures, where the progress of the participants is monitored at strategic checkpoints along the route. This may be where a bridleway or track crosses a road, in a similar way to where, in a canoeing venture a road crosses the river or at a weir or lock. Closer supervision may be necessary on occasions but it must always be sufficiently remote from the participants to avoid destroying the group's sense of self-reliance.

A horse box or trailer placed in a strategic location may be of considerable assistance in recovering a horse if it should go lame, lose a shoe or require veterinary care. Supervisors should know the location and telephone number of the local vet or blacksmith, as it may be possible for a blacksmith to re-shoe a horse or a vet to treat a minor injury at the camp site.

ASSESSMENT

The Assessor must be familiar with the contents of Chapter 6.4 - Assessment, briefed for the role and should be accredited where possible.

The role of the Assessor in riding ventures is little different from that of assessing other modes of travel. The Assessor should be approved by the Operating Authority at Bronze and Silver levesl. For all ventures in Wild Country Areas the Assessor should be accredited. Many Wild Country Panel Assessors are capable of assessing riding ventures as their principal task is to ensure that the requirements and conditions of the Expeditions Section have been fulfilled and does not involve any responsibility for the care of the horses. Assessment should not present any particular difficulty. Most of the journeying will take place on tracks, lanes and bridleways, and camp sites will rarely be located in places which do not have vehicular access because of the need to pre-position food, fodder and camping gear.

CONCLUSION

A clear identification of the task is a prime consideration in any venture, this is especially so in riding ventures. The advice in this chapter is intended to assist this process. Though probably, in terms of the mode of travel, the most demanding ventures in the Expeditions Section, an increasing number of young people are completing riding ventures. The sense of satisfaction and fulfilment following a successful riding venture is probably unrivalled.

J. Driscoll

Water Ventures

An Overview

Water ventures provide a way of increasing the scope of adventurous activity as well as extending access to a considerable part of the world. The Expeditions Section recognises three modes of travelling on water - canoeing, rowing and sailing - and, since they take the form of journeys by the participants' own physical effort, they are all very friendly to the environment. They cause no pollution or erosion of the river banks and are silent in operation. Water has the great advantage that it does not wear out, unlike many of our overused areas of countryside. Britain is well served by rivers, but access to the majority of them is difficult unless there is an historic right of passage or they have been turned into 'navigations'. More ventures are now taking place on the Continent where many of the problems of access we have in this country do not exist.

The Award Scheme encourages alternative modes of travel to enable those with canoeing, sailing and rowing craft to use their expertise and resources as an alternative to traditional ventures on foot, as well as encouraging those who have little or no experience of water activities to seek fresh challenges. The Award is all about providing new learning experiences and the excitement of new challenges.

Water ventures provide an opportunity for those with mobility problems to participate and, though the ventures are demanding, participants do not have to carry their equipment on their backs.

Participants who have experience in handling canoes, rowing boats, sailing dinghies or yachts can be ambitious in their planning. Groups with little or no experience should not be deterred from the challenge of water ventures but would be wise to limit their ambition. Participants with access to canoes or boats and suitable water on a long-term basis should consider the advantages of undertaking water ventures at Bronze, Silver and Gold levels, so that they have the opportunity to build up their skills and experience progressively in the same way as those who carry out their ventures on foot.

Tall ships and large sail training vessels may not be used for ventures within the Expeditions Section, however such journeys may fulfil the requirements of the Residential Project at Gold level. Yachts up to a maximum of about 10-11 metres (32-36 feet) and with up to 8 or 9 berths may be used for Expeditions. Most participants are nowhere near so ambitious and use sailing dinghies or keel boats, camping ashore each night. In this *Expedition Guide* popular terminology is used. 'Yacht' is used for a keelboat with cabins providing 'built-in' sleeping accommodation; the term 'keelboat' is used for a boat with a keel but which does not provide permanent sleeping accommodation, and the crew, as dinghy sailors, have to sleep ashore at night unless they bed down on the bottom boards.

It is in the nature of water recreation that there are many tens of thousands of boats (sailing and rowing) and canoes on and around our coasts and waterways which are rarely, if ever, used. If more Award Scheme participants are to be able to undertake water ventures, it is essential that Operating Authorities manage to gain access to some of this under-used equipment. Water ventures will not become a realistic option for the majority of young people until Operating Authorities are able to offer the same support and back up facilities as they do for ventures on foot.

Water ventures in western Europe are practical for young people, particularly from the South of England, as the costs and travel implications are no greater than travelling to Scotland for a venture on foot. The same efforts must be made in identifying suitable water and facilitating canoeing and dinghy sailing as have been made for ventures on foot, and a team of knowledgeable and enthusiastic Adult Leaders will be needed to bring this about. Every effort should be made to raise the profile of water ventures and Operating Authorities with suitable facilities might pool resources to offer Open Gold water Expeditions for those who have the necessary competence but are unable to form a viable group.

SUCCESSFUL WATER VENTURES

The keys to success in water ventures are a well-chosen purpose, the appropriate water, suitable craft and a sufficient number of the right companions. The participants' proximity to suitable water, and the craft available will play a decisive role in the mode of travel, but other factors need to be taken into account. **It is more in keeping with the spirit and philosophy of the Expeditions Section for a group to travel unaccompanied, be self-sufficient and dependent on their own resources on less demanding water with visits from their Supervisor and Assessor, than to be in estuaries, coastal waters or open sea where the presence of a Supervisor is mandatory.** Drascombe Luggers or Wayfarers, on waters similar to the tidal areas of the Norfolk Broads with their crews camping or bivouacking ashore at night, are more in keeping with the spirit of the Expeditions Section than an offshore passage in an 11 metre yacht with both the Supervisor and Assessor aboard with their RYA/DTp Yachtmaster Offshore Certificates.

It is more appropriate to be unaccompanied on suitable water

Maggie Willis

SELECT WATER WITHIN THE GROUP'S CAPABILITY

The Award Scheme regards all modes of travel as of equal merit, but it is in the group's own interests to select water and craft which give an excellent chance of bringing the venture to a successful conclusion. A canoeing or rowing venture on a Grade 1 or 2 river is less likely to be affected by the weather than a sailing venture on an estuary.

Rivers are subjected to occasional flood conditions, even during the summer months, but the water used by open canoeists and dinghy sailors is only occasionally affected by too little water. Those who sail may be subjected to either too much or too little wind which may make journeying difficult or impossible. On exposed water, strong winds create the additional problem of rough water which may also impede or prevent progress.

Water should be selected which is suitable for the craft involved and well within the level of skill of the group. Ventures will normally be unaccompanied and self-reliant unless the venture takes place in estuaries, sheltered coastal water or open sea. Participants must be able to train and carry out practice ventures on water of a similar degree of difficulty so that they become proficient and experienced in coping with any problems which may arise. Where the water is within the participants' competence, there is less chance of adverse weather or environmental conditions affecting the journey, which ensures a more predictable outcome to the venture.

FURTHER ADVICE

Having decided on suitable craft and located the appropriate water, the following advice should be considered when planning and preparing for water ventures:

- Allow more time to plan, train and prepare than would be necessary for a land-based venture unless the previous qualifying venture used the same mode of travel.
- Incorporate greater flexibility into the planning and execution of the venture.
- Build alternative back-up dates into the planning in case of unsuitable weather.
- Involve experienced, adaptable Supervisors and Assessors.

More Time to Plan and Prepare

In groups where all the members have had previous experience in their chosen mode of travel and have access to suitable craft, the time taken to plan, train and prepare will be very much the same as for a venture on foot. If the participants have little or no experience in the chosen mode of travel and are having to learn new skills from scratch, more time must be allowed. Where the mode of travel at Silver or Gold level is different from that at the previous level, extra practice journeys will be needed.

Incorporate Greater Flexibility into Planning and Execution

Weather affects all ventures on land or water, but the effect on water ventures is greater and more immediate. The impact of a headwind, no wind or a strong current may decrease or increase the distance travelled by a considerable amount, so it is essential to plan with this in mind. Where ventures take place on inland waters, alternative camp sites should be considered to accommodate the varying distances travelled. Planned starting and finishing times may have to be changed at the last minute to meet changes in water or weather conditions. If this is the case, the Supervisor and Assessor must be informed of the changes.

Allow more flexibility for water conditions

D.M.Boyes

Build Alternative Back-up Dates into the Planning

Even with the greatest flexibility in planning and execution, there comes a time when a venture cannot take place or has to come to a premature end because of weather or water conditions. A river in spate or reduced to a trickle, gale force winds or no wind at all, may prevent the activity taking place. All who engage in outdoor pursuits or activities have to come to terms with this reality sooner or later. There has to be a philosophical acceptance of this in return for all the pleasures and excitement the pursuits engender. This should be taken into account, particularly when participants are nearing the upper age limit of their 25th birthday. At least one back-up date should be arranged for all qualifying ventures on water as the probability of ventures having to be postponed or aborted is greater than on land. More than one back-up date should be considered for Expeditions on estuaries, coastal waters and the open sea.

Involve Experienced, Adaptable Supervisors and Assessors

All the flexibility on the participants' part and several back-up dates are of no avail unless the Supervisors and Assessors can adapt to the needs of the group and their venture. Experienced Supervisors and Assessors who may be available for both the target date and the back-up date(s) must be identified at a very early stage in the planning. Supervisors and Assessors who have the flexibility to meet last-minute changes to the schedule and route have an important role in water ventures. It is very difficult to assess a water venture if the task clashes with other commitments, and Supervisors and Assessors should be careful when offering their services if they have doubts about their availability. The option to change plans and routes has always been available to land-based ventures to meet weather and safety demands. This flexibility is available for ventures on water providing that the group, Supervisor and Assessor are all involved in the discussions and decision-making. **It is not permissible for an Expedition to be changed into an Exploration simply to comply with the travelling requirements.**

Local Pre-expedition Checks are probably even more important for water ventures than they are for ventures in wild country, especially where craft are being trailed to distant locations. Local Pre-expedition Checks are advisable at Bronze, Silver and Gold levels to ensure that equipment and training is to the required standard.

Following this advice should greatly enhance the chances of success. The rewards and satisfaction which arise from the successful completion of a venture on water are so great that they will more than repay the extra planning and effort which they frequently entail.

SAFETY AFLOAT

Water is a very exciting and challenging environment, so it is inevitable that it will also be a very demanding one. The mountain scene can vary rapidly but it never changes as quickly as conditions on water. Situations can change with frightening rapidity. Safety must always be a vital consideration for all who engage in ventures on water, or people responsible for caring for others. The Common Training Syllabus applies to all ventures which take place on water. **Particular attention should be paid to those aspects of basic first aid concerned with resuscitation and the ability to recognise and treat hypothermia.** Resuscitation is included in the syllabus for all ventures at Bronze level but not the recognition and treatment of hypothermia; for water ventures this must be included.

Participants should be able to administer resuscitation on and in the water, as well as adjacent to it. Because of water's ability to drain heat from the body and the comparatively low temperature of the water around the British Isles at all seasons of the year, the probability of succumbing to hypothermia is greater on water than on land. Hot weather brings its own problems. Participants need to be aware that prolonged exposure to the sun on water can lead to illness. Every year many hundreds of thousands of people find satisfaction and enjoyment from water recreation without coming to the slightest harm, and a few simple precautions and procedures will ensure the safety of all concerned. The safety of Award ventures afloat depends on the following factors:

- **The ability to swim at least 25 metres without any buoyancy aid and in light clothing.**
- **A buoyancy aid or lifejacket and suitable clothing and footwear must be worn.**
- **A craft with adequate buoyancy.**
- **A practised and predictable response to sudden immersion.**
- **Proficiency in capsize and recovery drills, or man overboard drill.**
- **Appropriate qualifications or equivalent levels of competence.**

- The ability to assist each other in difficulty.
- Training for an Expedition and not just in the technical skills.

Swimming at least 25 Metres without a Buoyancy Aid

The Award has changed the old requirement of swimming 50 metres, with or without a buoyancy aid, to an unequivocal requirement to **swim 25 metres without any buoyancy aid and in light clothing**. The ability to swim 25 metres is an expectation of Keystage 2 of the National Curriculum. It is intended for the eleven year old, but it will serve as a foundation on which to build water safety. Swimming 25 metres is an indicator of water confidence.

The Buoyancy Aid or Life-jacket and Suitable Clothing and Footwear

Situations arise where the ability to swim is not sufficient in itself to ensure safety. All participants must wear a buoyancy aid or life-jacket when on or close to the water with specific exceptions which are dealt with in Chapter 4.7 - Rowing. The type of buoyancy will be determined

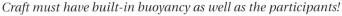

Craft must have built-in buoyancy as well as the participants!

by current custom and practice relating to the type of craft being used. Inflatable life-jackets are notoriously difficult to keep in good order. Do not trust anyone else with your safety. There is only one way of ensuring that buoyancy aids and life-jackets are sound - and that is to test them.

Water, far more than wind, drains heat from our bodies so hypothermia is always a concern for those who engage in water ventures in and around the British Isles at any time of the year, whether one is soaked on top of the water or immersed in it. Suitable dress for a particular activity should be determined by current good practice. Clothing must give protection during the activity and when immersed in water, and yet must not hamper movement. It is not always necessary to wear wet suits, though they do provide very good protection. Traditional clothing can be very effective in reducing heat loss when immersed in water and modern synthetic materials give very good insulation without absorbing as much water as the more traditional materials.

Protection of the extremities, head, hands and feet is most important; the head in particular can be a great source of heat loss. A thermal hat such as a thick knitted woolly bobble cap provides cheap and effective protection for the head, especially when supported by a waterproof anorak hood. Buoyancy aids and life-jackets provide high visibility in the event of an emergency; brightly coloured headwear or upper body clothing can add to this visibility. Gloves, or paddle mitts, are a great comfort against blisters and the cold when canoeing. Protection from the sun is very important in sunny weather, especially as light is also reflected from the water. It is usually difficult to change position or posture, and frequently impossible to find shade; severe burning may therefore occur if the skin is not adequately protected, either by clothing or blocking creams. Feet should always be protected; deck shoes, plastic sandals or old trainers are a necessity for all water ventures to avoid cuts from broken glass, abrasions and the possibility of infection. Wet suit bootees are not satisfactory for river ventures unless an over-shoe is also worn.

A Craft with Adequate Buoyancy

All craft should have sufficient built-in buoyancy to stay afloat and support all the occupants. If the buoyancy is not built into the craft, then it must be firmly secured. As with life-jackets, this should never be taken on trust - even the best regulated establishments have been known to fall down on this vital provision - the buoyancy of all borrowed or hired craft should be tested by the users. In the event of a capsize the golden rule is 'always stay with the boat'. It is much more visible than a head bobbing up and down in the water.

A Practised and Predictable Response to Sudden Immersion

Sudden immersion in cold water is an inevitable consequence of taking part in water activities. All training must take place with this in mind. The effect of sudden immersion may result in unpredictable responses, even from the young and healthy. Occasionally heartbeat irregularities occur with the possibility of dire consequences. Hyperventilation occurs and nearly always there is an involuntary gasping for breath which may lead to the ingestion of large quantities of water. It would be most unusual for people to be able to hold their breath for more than twenty seconds in cold water. Suitable clothing may go a considerable way to reducing this shock. This should not discourage anyone who wishes to participate in water activities, but **it is essential that all participants know how they will react if they are suddenly immersed in cold water.** Thousands of people fall into water everyday during the summer without any ill effects. It is possible, by progressive training, to become accustomed to falling into cold water and lose one's fear of sudden immersion.

Confidence on the water can be greatly increased by being at home in and under the water. In the Service Section of the Award there are the life saving qualifications, while the Physical Recreation Section has relevant qualifications in swimming and personal survival. Any of these would be most helpful in establishing personal competence and confidence in the water and would increase the ability to help other participants and the public in general. These qualifications make excellent supporting choices for all those engaged in the Award who wish to carry out water ventures.

Proficiency in Capsize and Recovery Drills

The importance of these drills is obvious and yet they are frequently neglected because of the reluctance on the part of many novices to get wet because of the British climate and the low ambient temperature of water in and around the British Isles. When it is too cold for the participants to remain soaked for long, an intensive period of capsize and recovery drill should always come at the end of the training session. The operative word is 'drill'. Capsize should always bring about an

A predictable response to sudden immersion

automatic reflex response which does not involve conscious thought; only then can one be confident that, no matter how great the anxiety or stress, panic will not ensue. **Capsize must be followed by recovery** which may range from swimming or dragging a canoe to the bank, or by using one of the deep water rescue techniques in the BCU training syllabus. Similarly there is an RYA training procedure for righting and emptying a capsized dinghy. On yachts and keelboats this should be replaced by 'man overboard' drill. The end of a training session should provide an opportunity to practise these drills and to test the ability to swim 25 metres out of doors (not in a heated swimming pool) without a buoyancy aid.

Have the Relevant Qualifications or Equivalent Competence or Experience

The Royal Yachting Association and The British Canoe Union have appropriate qualifications and syllabuses for sailing and canoeing ventures in the Expeditions Section. The level of competence required is listed in the chapters concerned with the various modes of travel on water at Bronze, Silver and Gold levels. For rowing ventures, where there is no appropriate national qualification, a syllabus has been prepared and is included in Chapter 4.7 - Rowing. The Sea Cadets, Scouts, Guides and some Operating Authorities have their own courses of instruction. Training must be supported by experience. It is important that regular practice is spread over a period of time, for it is all too easy for participants to give the appearance of competence and respond in the correct manner, yet still lack the depth of experience and confidence to cope with an emergency.

Being Able to Assist Each Other when in Difficulty

The Expeditions Section is all about groups working together for a common purpose, and the efforts of each individual is essential for the success of all. Participants depend on each other and should have the requisite skills to help other members of the group, including the ability to throw a lifeline accurately, right an upturned boat and administer resuscitation. The skills related to the activity and to survival are vital, but they are not sufficient in themselves. Confidence in each other can only come about through training and practising together over a considerable period of time until the individual members become welded into a team - which takes us back to the beginning: plan, train and prepare for water ventures a long way ahead so that there is adequate time to become an integrated group sharing a mutual confidence in each other's skill.

Train for an Expedition - not just the Technical Skills

Training should be directed towards the completion of a journey on water and not just the mastering of a number of technical skills related to handling a particular craft. Technical skills, though vital, are a means to an end, not an end in themselves. The training must lead to an awareness of the environment in which the venture takes place and any potential threats or unexpected hazards which the surroundings may present, whether this involves the removal of fish hook barbs or the ever-present water pollution which is always a problem on our rivers and coasts.

Weil's disease (leptospirosis) is a bacterial infection carried in rat urine and can cause serious illness leading to kidney or liver failure. The illness is a **notifiable disease** which requires hospital treatment. The bacteria contaminate river and canal banks and water, particularly stagnant or slow moving water. It is also present in other inland water, such as small lakes, with higher concentrations at the edges. The early symptoms are a high temperature and muscular pain - influenza-like symptoms. If these symptoms are experienced after any water activity a doctor should be contacted immediately and told of the activity and the concern. As a general precaution, all participants should ensure that any open cuts are protected by waterproof adhesive dressings. Another potential hazard on some still, inland waters is the presence of blue-green algae, which can cause painful skin irritation after immersion. If the risks from either of these hazards are high, participants may find particular stretches of water are closed for the use of watersports. The hazards in water ventures are no greater than for land-based ventures - they are just different! Participants who have been properly trained by knowledgeable and experienced Instructors are in no greater danger than those engaged in land ventures.

For canoeing, sailing and rowing ventures which take place on rivers and canals, the standard approach to navigation using the syllabus set out in the *Award Handbook*, is more than adequate. Only when tidal waters, estuaries and coastal waters are being used is it necessary to use charts and change to marine navigation.

THE USE OF MOBILE PHONES

Mobile phones have a very restricted use in Wild Country Areas because of the screening effect of the terrain, but their use is much more predictable in coastal areas and their range frequently extends over wide estuaries and well out to sea. Mobile phones usually have good coverage in the open valleys used in river ventures. They provide Supervisors and Assessors with a very effective means of communication and, on sheltered estuaries and coastal waters, they may offer the advantages of 'ship to shore' communication without the need to obtain a Wireless Operators' Certificate. Equipment should be tested in the area of the venture prior to the qualifying venture. Telephones must be protected from immersion in the water and secured to the craft or to the participants. For more detailed information refer to Chapter 6.3 - Mobile Phones and Radios.

Maggie Willis

Canoeing

This chapter must be read in conjunction with
Chapter 4.4 - Water Ventures - An Overview.

Canoes fall into two distinct categories - the kayak and the open canoe. When speaking of the activity and the craft in general, the word 'canoe' will be used in this chapter, except where it is necessary to refer specifically to one category or the other.

Canoes are probably the cheapest and most readily available method of carrying out a venture on water. There are many thousands of them in the British Isles, most of them under-used. It should not be beyond the ingenuity of Award participants to borrow canoes if they do not possess their own; failing this, they might be able to hire them at a modest cost. A journey by canoe gives an entirely new perspective on journeying, and to the development of new skills and experience frequently leading young people into an activity which may last a lifetime; yet the levels of skill and experience may realistically be acquired in the time it takes to gain either a Bronze, Silver or Gold Award.

Canoeing ventures on inland waters are no more likely to be affected by foul weather than ventures on foot and there is an excellent chance of a predictable and successful outcome to such ventures.

Canoes lend themselves very readily to Explorations and provide an opportunity to engage in fresh areas of study in a stimulating environment; they also offer an exciting form of travel which can lead to new interests and awareness.

Maggie Willis

Licences are required for all English and Scottish canals

WATER

Unlike in continental Europe and most other countries of the world, access to rivers in the United Kingdom is difficult, and frequently impossible, except where they have been turned into navigations or there is a tradition of access. Navigations usually involve the lower reaches of rivers which, in many cases, are still used for commerce and recreational boating. While they do not provide excitement for the white water enthusiasts in their kayaks, they are ideal for those engaged in Award ventures transporting their camping gear and equipment. They include the Severn, the Wye, the Yorkshire Ouse, the Great Ouse, the Thames and the Trent, along with many others, with their British Canoe Union (BCU) Grade 1 and 2 water, though at Gold level one is looking for more than 128 kilometres, or 80 miles, of suitable water.

Canoeists require licences for all English and Scottish canals and some canalised rivers. The British Canoe Union has negotiated a special arrangement with the British Waterways Board whereby their waters are included in the BCU Licence, which is free to BCU members. Separate licences are required for many of the larger rivers, such as the Thames or the Great Ouse. The relevant details may be obtained from the BCU.

On the Continent there are thousands of miles of suitable rivers used by tens of thousands of touring canoeists on a scale difficult to imagine in Britain. Most have their source in the Alps or the Massif Central. While their upper reaches have no place in the Expeditions Section and their lower reaches are frequently polluted and highly commercialised, the middle sections, often extending for hundreds of miles, are a delight for canoeists. The Dordogne, the Ardèche, the Rhine, the Rhône and the Danube are typical examples and provide wonderful, stimulating environments for canoeing Expeditions and Explorations. Some rivers have been canalised with locks, and licences may be required.

THE CRAFT

The **open canoe** with its origins in the Canadian canoes of the North American Indians and the Voyageurs, is an ideal craft for Expeditions. It is an excellent vehicle for carrying equipment and has that little bit extra space which may be needed for an Exploration. They allow for companionship, usually carrying two or three people, and are safe insofar as if you fall in, you fall out; there is little danger of being trapped by the coaming or spray-deck as in a kayak and therefore novices do not have the same problem in mastering the capsize and recovery drill. Not being designed for use in estuaries or coastal waters, open canoes are most suited for use on rivers; with care and judgement they can be used by suitably experienced people on lakes and lochs, as their prototypes were in Canada. When faced with an open expanse of water in windy conditions, it is necessary to keep to the windward side of the lake under the shelter of the shore where the water is calmer; in the middle or on the lee shore the waves have a longer fetch and may swamp an open canoe.

The modern, low volume **white water kayak**, though providing excellent sport and used in vast numbers throughout the country, is not a very suitable craft in which to carry out a journey as there is rarely enough room to carry camping gear and food. Many young people find them difficult and tiresome to paddle in a straight line for five or six hours at a stretch. Outdoor Centres use these kayaks in large numbers and thousands of young people enjoy their first taste of canoeing in them. They are exciting and very good for training and mastering kayaking skills but, as far as the Award Scheme is concerned, they are more suited to the Physical Recreation Section.

The **sea kayak** is an excellent craft in which to undertake a camping venture. With their length and keel they are easy to paddle in a straight line and there is ample room for camping equipment. Though designed for the sea, they are suitable for use on the type of river involved in Award Expeditions, and canoeists who camp and travel long distances prefer them. There are also an increasing number of purpose-built **touring kayaks** becoming available which are ideal for Award ventures.

REQUIREMENTS

The requirements for canoeing ventures are the same as those for all other ventures. See Chapter 2.1 - Introduction and Requirements.

CONDITIONS

Participants using inland waters must observe the Country Code.

All participants must know:
- The Water Sports Code.
- The basic rules of the water - priorities, the sound signals used on water and distress signals.
- The courtesies, customs and etiquette associated with boating and sailing.

All participants must be adequately trained to:
- Demonstrate that their equipment is waterproofed.
- Satisfy the Assessor that their kayak or canoe, equipment and clothing is suitable for the venture.
- Satisfy the Assessor as to their competence. The levels of competence are given below. Participants must hold the following qualifications or be of a equivalent level of competence.

It is essential that all Instructors have access to the Directory of Tests and Awards published by the Coaching Service on The British Canoe Union website: www.bcu.org.uk

BRONZE

Duration:	A minimum of four hours of paddling time on each of the two days.
Water:	Rivers, canals, lakes and other inland waters in rural areas.

Competence: Open Canoe BCU Open Canoe 2 Star Test.

              Kayak   BCU Closed Cockpit Kayak 2 Star Test.

                                   BCU Placid Water 2 Star Test.

There must be a minimum of two craft to render mutual support.

There is no requirement concerning distance, but a reasonable expectation would be for a group to cover 16-19 kilometres, or 10-12 miles, each day, depending on the number of portages and weather-related factors such as headwinds. There is no requirement that the venture should take place on moving water. Canals and rivers up to BCU Grade 1 are quite suitable and the water may be familiar to the participants.

SILVER

Duration: A minimum of five hours of paddling time on each of the three days.

Water: Rivers, canals and other inland waters in rural areas. The water must present an appropriate challenge and must be **unfamiliar** to the participants.

Competence: Open Canoe BCU Open Canoe 2 Star Test.

Kayak BCU Closed Cockpit Kayak 2 Star Test.

BCU Placid Water 2 Star Test.

There must be a minimum of two craft to render mutual support.

Canals are acceptable, as are Grade I rivers, and there is no requirement for the water to be moving. There is an expectation, however, that the conditions will be related to the age and experience of the participants and represent a progression between the Bronze and Gold levels of the Award. Venturers might be expected to journey some 72-77 kilometres, or 45-48 miles, depending on the number of portages and weather related factors such as headwinds.

GOLD

Duration: A minimum of six hours of paddling time on each of the four days.

Water: Rivers, navigations, lakes, lochs and sheltered coastal waters may be used. The water must be **unfamiliar** to the participants and must present an appropriate challenge.

Competence: Open Canoe BCU Open Canoe 3 Star Test.

Kayak BCU Closed Cockpit Kayak 3 Star Test.

BCU Placid Water 3 Star Test.

There must be a minimum of two craft to render mutual support.

Participants intending to venture on coastal waters should be trained and have passed the Sea Kayak 4 Star Test.

A portage round a lock or weir usually takes 20-30 minutes

At Gold level the water must be moving, either with the current or tide, but there is no requirement that the water should exceed BCU Grade 2. The current should not be in excess of 4-6.5 kilometres (3-4 miles) per hour. The Award strongly disapproves of using water of greater difficulty in ventures within the Expeditions Section. Venturers might be expected to journey a little over 32 kilometres (20 miles) per day depending on the strength of the current, the number of portages or headwinds resulting in a journey of some 130-145 km or 80-90 miles.

EQUIPMENT

All craft must be sound, suitable and fitted out for the conditions in which they are to be used. They must have built-in buoyancy, or buoyancy which is securely attached to the boat. The buoyancy should be sufficient to support the canoe and its occupant(s) and there should always be a practical buoyancy test. Kayaks should have well-fitting spray decks.

General Equipment Related to the Boat

- Throwing line.
- Spare paddle.
- Bow and stern toggles, or loops, and buoyant painters fore and aft, properly secured.
- Repair kit including canoe tape, very thick flexible plastic sheeting etc.
- A bailer and large sponge (open canoes).
- Maps/charts in water resistant protection.

Canoeists using sheltered estuaries and sheltered coastal waters should also carry the following:

- Compass(es).
- A waterproofed/water resistant chart.
- A powerful water-resistant torch.
- A towline (instead of a throwing-line).
- Flares.
- Emergency rations and water.

Personal and Expedition Equipment

Participants should refer to the equipment list in Chapter 3.1. Most equipment is the same for all modes of travel with adjustments being made where necessary.

Clothing

The list of personal equipment in Chapter 3.1 may be used with the necessary substitutions. Additional suitable clothing will range from wet or dry suits, providing protection against the cold, to light coverings giving shelter from the sun. Protection for the head, hands and feet is particularly important. Feet must be protected at all times; deck shoes, plastic sandals or even old trainers may be used. A complete change of clothing should be carried in a waterproof container.

Personal and Emergency Equipment

This list has been amended to make it more relevant to the canoeist's needs:

- Buoyancy aid or life-jacket as appropriate, with whistle attached.
- Bivvy bag (poly-bag).
- Waterproof or water-resistant watch.
- Sailor's knife with blade, tin-opener and spike.

- Matches in waterproof container.
- Personal first aid kit.
- Emergency rations.
- Small water-resistant pocket torch with spare bulb and alkaline batteries.
- Notebook and pencil.
- Coins/phonecard for telephone.

Personal Camping Equipment

This equipment list is virtually the same for land and water ventures with the substitution of waterproof containers for the rucksack.

Group Camping Gear

Similarly, group camping gear is practically the same for all ventures. The weight and bulk of camping gear and food should be kept to a minimum when canoeing. Canoeing involves portages not only at the beginning and end of the day, but frequently during the journey itself around weirs, locks and some rapids. British Waterways do not allow canoeists to use locks, unlike the National Rivers Authority, Thames Region. Portage can be a strenuous activity, especially for younger participants, and frequently involves several trips between the place of disembarkation and re-embarkation. Portage round a lock or weir will usually take 20-30 minutes.

WATERPROOFING EQUIPMENT

All clothing, much of the camping gear and food will need to be protected in waterproof containers or dry bags. If there is insufficient room in the waterproof containers, stoves, pots and pans will come to no harm in a string bag provided it is secured to the canoe. Large plastic drums with an efficient seal are very popular with open canoeists, while kayakists tend to use dry bags or smaller plastic drums. These plastic drums were originally designed to hold chemicals but are now available on the open market. Waterproof bags inside rucksacks are an alternative, being particularly useful at portages and adjusting the trim of the canoe. Some people dislike large plastic drums because of the difficulty of handling them at portages and only use small drums. Whatever method of waterproofing is used, it must be effective. Spare clothing and sleeping bags should be given the additional protection of being individually wrapped in heavy-gauge plastic bags sealed with elastic bands. All equipment, including those two vital items, the spare paddle and the throwing-line, must be securely attached to the canoe so that

Clothing and gear must be protected in waterproof containers or dry bags

everything stays together in the event of a capsize, though many canoeists prefer to have the throwing-line on their person.

In canoes the storage of equipment affects the trim and stability of the vessel. Equipment should be stored to ensure that the canoe is appropriately trimmed for the prevailing conditions, and heavy items stored as low as possible to increase stability.

TRAINING

The Common Training Syllabus as set out in the *Award Handbook* should be followed in conjunction with the advice in Chapter 4.4 - Water Ventures - An Overview. The land navigation syllabus will usually suffice for ventures on inland waters; elsewhere marine navigation should be substituted for land navigation. The training related to canoe and kayak skills should be at the level listed earlier in this chapter. **Reference must be made to the relevant BCU syllabuses and any additional Operating Authority requirements.** Participants should be trained to use the repair kit and carry out emergency repairs.

There is no substitute for experience and days should be spent on the water until the canoe becomes an extension of oneself and the canoeist is at home in, on and under the canoe. When the basic skills have been acquired, it is essential that experience is built up using the same kind of water that is to be used for the qualifying venture. This is particularly important for ventures on rivers and coastal waters where it is essential that participants are able to 'read the water'. Supportive skills, such as self and group rescue techniques, must be practised until they are automatic responses. Portages and handling the canoes in and out of the water in difficult situations must become efficient routines. It is vital that the group is trained to keep together to ensure safety in numbers. With all canoes, unless there is danger of drifting down to some hazard such as a weir, canoeists should hang on to the canoe and either swim with it to the bank or await rescue. Canoeists and kayakists should always employ the capsize drills and self and group rescue techniques learned in their training.

Canoeists must be as proficient in the skills of map-reading as those who venture on foot so that they always know where they are on the river, and are able to anticipate any hazards that the river may have in store. **Participants on Award ventures must stay away from weirs and their attendant stoppers**. Groups using estuaries and sheltered coastal waters will have to master the seamanship associated with tides and currents in addition to the ability to use charts and tide tables. See Chapter 4.7 - Rowing (Boatwork Syllabus Part C - Tidal) for further details.

WHERE TO CARRY OUT THE VENTURE

The choice of where a venture is carried out will depend to a large extent on its purpose, but choice will be assisted by referring to the British Canoe Union's literature. Their River Guides provide detailed advice on the rivers as well as details of access and rights of passage. The BCU will also be able to give advice on many of the European Rivers, or tell you where information is available. The address of the BCU is listed in the Appendix along with that of The British Waterways Board who will provide the regional address from where canal licences may be obtained. Licences are required for rivers such as the Thames and the Norfolk Broads which are maintained as navigations. Membership of the BCU can include a British Waterways Board Licence.

Participants should choose water which is suitable for the purpose of the venture and the type of canoe and, above all else, water which is well

Maggie Willis

The canoeist must be at home in, on and under the canoe

within the competence and experience of the participants. One of the principal conditions of the Expeditions Section is that all ventures should be self-reliant and unaccompanied. An exception has to be made for canoeing ventures which take place in estuaries and coastal waters where, for reasons of safety, they have to be supervised on the water by a suitably experienced adult. Again, it is more in keeping with the spirit of the Award for a group of suitably trained and experienced Gold participants to carry out an unaccompanied, self-reliant Expedition on a Grade 1 or 2 river, than to engage in a coastal venture where they have to be accompanied by an adult.

SUPERVISION

The Supervisor must be familiar with the contents of Chapter 6.2 - Supervision and be approved by the Operating Authority.

The British Canoe Union offers appropriate qualifications and levels of experience for Supervisors. The BCU Level 2 Coach, Level 3 Coach, Placid Water Level 3 Coach (Inland or Sea as appropriate) are suitable qualifications, or the equivalent level of experience for ventures on inland waters. Supervisors must have considerable experience of water similar to that being used, be competent in assessing water and weather conditions and must be approved by their Operating Authority. Canoeing ventures which take place on rivers, canals and navigations can usually be supervised from the bank providing the Supervisor has road transport and is able to meet the group at suitable locations during the day and at camp sites. Supervisors of river and canal ventures must always be within easy reach of the participants and able to respond within a reasonable period of time.

Where ventures take place in sheltered estuaries or on sheltered coastal waters, the Supervisor must provide safety cover afloat. Contact should not be made with the group during the journey except for the needs of supervision. The safety cover must be sufficiently remote from the participants to avoid destroying the group's sense of remoteness and self-reliance and yet be able to render assistance in an emergency within a reasonable amount of time. The safety cover may consist of the Supervisor and the Assessor, who will be appropriately qualified and experienced to at least Level 3 Coach (Sea), shadowing the group at an appropriate distance by canoe or safety boat. Where a safety boat is used it must be sufficiently seaworthy and fitted-out to cope with any water conditions which may arise in the sea area being used. It is desirable that all canoes used on open water should be highly visible in bright colours such as red, orange or yellow.

Written approval should be obtained from the Operating Authority for the venture and its supervision.

ASSESSMENT

The Assessor must be familiar with the contents of Chapter 6.4 - Assessment, briefed for the role and should be accredited if possible.

It is essential that those who assess canoeing ventures have canoeing experience and are able to evaluate canoeing competence and water conditions. Assessors should have the same level of qualification, or equivalent experience as the Supervisor - BCU Level 2 Coach, Level 3 Coach or Placid Water Level 3 Coach (Inland or Sea as appropriate).

Only the Severn and Wye Panel is dedicated to assessing canoeing ventures at Gold level (see Appendix). However, many of the Wild Country Panels have expert canoeists among their members who are qualified and willing to assess canoeing ventures.

For ventures in sheltered estuaries or sheltered coastal water, where the venture is accompanied by a safety boat, Assessors may also undertake the role of Supervisor providing they have the necessary qualifications, or equivalent levels of experience and have the approval of the Operating Authority (see above). They should be totally independent of the Award Unit at Silver level. At Gold level they **must** be totally independent of the Award Unit and not associated with it in any way. Where one person takes on the dual role of Supervisor and Assessor it halves the amount of adult intervention in the journey and enhances the participants' feeling of self-reliance. Many Supervisors will, however, appreciate the support of Assessors in situations which can be very lonely.

Assessors need to be flexible to cope with changes of plan and should have sufficient time at their disposal to enable them to adapt to the needs of the venture. As with all ventures, there should be close liaison with the group and the Supervisor before and at all times during the venture.

Sailing

This Chapter must be read in conjunction with
Chapter 4.4 - Water Ventures - An Overview.

Sailing ventures fall into two main categories which are so different in their nature and form that it is easier to consider them separately:

- Those using sailing dinghies, such as the Wayfarer, or open 'day-sailing' keelboats, which usually take place on rivers, sheltered estuaries, lochs and lakes, and where the participants camp or bivouac ashore each evening.

- Those which take place in yachts up to a maximum of about 10-11 metres in length (32-36 feet) with berths for up to 8-9 people, usually in estuaries, coastal waters or involving making a passage on the open sea, and where the participants live and sleep aboard for the duration of their venture.

As mentioned previously, tall ships and sail-training vessels may not be used for the Expeditions Section. Those wishing to undertake a venture on the water must decide from the outset which kind of craft they are going to use as all the training and preparation will be determined by their choice of craft and the appropriate water.

The training, skills and competence must be related to the type of craft and the mandatory practice journeys must take place in a similar type of vessel.

VENTURES IN SAILING DINGHIES AND OPEN KEELBOATS

Expeditions or Explorations in dinghies offer an exciting alternative to ventures on foot for, while the boats are relatively expensive to purchase, there are many thousands in use all over the country in outdoor centres and clubs, and there are tens of thousands of young people who have access to these craft and are already competent in their use. They have the advantage of having great mobility, being easily trailed behind a car and transported to suitable water. They do not require expensive ancillary equipment such as two-way radios, and maintenance is relatively cheap and easy when compared with yachts. Because of their widespread use it may be possible to obtain access to dinghies through a Local Education Authority or a National Voluntary Youth Organisation.

REQUIREMENTS

The requirements for sailing dinghy and keelboat ventures are identical with those for all other ventures. See Chapter 2.1 - Introduction and Requirements.

CONDITIONS

In order to encourage participants to become involved in sailing ventures the Award Scheme has changed the conditions relating to their duration. The requirement of six, seven and eight hours sailing at Bronze, Silver and Gold levels respectively has now been changed to an average of six, seven and eight hours of planned activity each day. This will enable the participants to include the time taken to stow equipment, rig and launch their craft in the morning and the opposite process at the end of the day. Reducing the sailing hours should widen the choice of water available to the participants.

BRONZE

Duration:	Twelve hours of planned activity over the two days, averaging six hours each day.
Water:	Rivers, canals, lakes and other inland waters.
Competence:	All participants must attain the RYA National Dinghy Certificate Level 3 or have an equivalent level of competence.
	Sailing dinghies must be used.

There must be a minimum of two craft to render mutual support.

SILVER

Duration:	Twenty one hours of planned activity over the three days, averaging seven hours each day.
Water:	Inland waters or sheltered estuaries which present an appropriate challenge and are **unfamiliar** to the participants.
Competence:	All participants must attain the RYA National Dinghy or Keelboat Certificate Level 3 or have an equivalent level of competence.

There must be a minimum of two craft to render mutual support.

GOLD

Duration:	Thirty two hours of planned activity over the four days, averaging eight hours each day.
Water:	Inland waters such as the tidal areas of the Norfolk Broads, rivers, lochs, lakes, estuaries and sheltered coastal waters may be used. The surroundings must present an appropriate challenge and be unfamiliar to the participants.
Competence:	All participants must attain RYA National Dinghy Certificate Level 5 or RYA Day Skipper Shorebased and Practical Certificates or have an equivalent level of competence.

There must be a minimum of two craft to render mutual support.

EQUIPMENT

All craft must be sound, suitable and fitted out for the conditions in which they are to be used. Dinghies must be fitted with adequate buoyancy which is securely attached to the boat and their buoyancy should be tested.

While participants are spared the necessity of having to carry food and equipment on their backs, the same discipline in restricting the weight, bulk and number of items must be exercised in sailing ventures. There is, however, more scope for carrying recording equipment or apparatus associated with an Exploration. Canned foods, always a standby for sailors and which do not require waterproofing, are frequently carried but they must be identified with a spirit marker in case the labels wash off. With the exception of such items as pots, pans, stoves, fuel and canned foods which only need to be secured to the boat in a string bag to

Those using keelboats will have to sleep on the bottom boards or ashore

keep them ship-shape, all food and equipment must be waterproofed in dry bags or plastic drums. Sleeping bags, spare clothing and cameras should be given extra protection by being individually wrapped and sealed in plastic bags. All containers, as well as ancillary equipment such as paddles, bailers and anchors, must be securely attached to the boat.

Many dinghy sailors prefer to bivouac under a plastic sheet or boom cover rather than use a tent. Do not use the mainsail as a cover, except in life threatening circumstances, as there is a real danger of ruining an expensive sail by chaffing.

It is impracticable to carry a launching trolley on the venture but two or three lengths of plastic drainpipe,each about one metre long, will make helpful rollers and go some way to replacing the trolley and ease the handling of a dinghy in and out of the water. Offcuts from a builder's yard will serve very well. A loaded dinghy can be a handful for two, three or even four young people at the best of times, especially when it has to be carried above the high water mark. It is expected that open keelboats will be moored or securely anchored overnight.

The ancillary equipment will be determined to some extent by the water being used but the following list will serve as the basis:

General Equipment Related to the Boat

- Anchor and warp.
- Throwing line.
- Painters and/or spare line.
- Bucket, bailer/bilge pump and sponge.
- Charts/maps in water resistant protection.
- Compass(es).*
- Repair kit.
- Buoyant knife.
- Powerful water-resistant torch. *
- Flares, air horn.*
- Emergency water supply.*
  * *These may not be necessary for river sailing.*

Participants should also refer to the equipment list in of Chapter 3. 1. Most equipment is the same for all modes of travel with adjustments being made where necessary.

Clothing

This list may be used as a checklist with the necessary substitutions - but remember that it can become very cold around the camp site at night:
- Suitable clothing including protection for the head, hands and feet.
- A complete change of clothing in a waterproof container.
- Protection for the feet must be worn at all times - dinghy boots, plimsolls, deck shoes or plastic sandals are ideal.

Personal and Emergency Equipment

This list has been amended to make it more relevant to the sailors' needs:
- Life-jacket or buoyancy aid with attached whistle.
- Bivvy bag (poly-bag).
- Waterproof or water-resistant watch.
- Sailor's knife with blade, tin-opener and spike.
- Matches in waterproof container.
- Personal first aid kit.
- Emergency rations.

Many Operating Authorities have access to dinghies

- Small water-resistant pocket torch with spare bulb and alkaline batteries.
- Notebook and pencil.
- Coins/phonecard for the telephone.

Personal Camping Equipment

This equipment list is virtually the same as for land ventures with the substitution of waterproof containers for the rucksack.

Group Camping Gear

Group camping gear is practically the same for both land and water ventures.

TRAINING

The Common Training Syllabus as set out in the *Award Handbook* should be followed in conjunction with the advice given in Chapter 4.4 - Water Ventures - An Overview. The land navigation syllabus will usually suffice for ventures on inland waters; elsewhere marine navigation should be substituted for land navigation. The training related to sailing skills should be at the level listed above and participants should be trained to use the repair kit and carry out emergency repairs.

SUPERVISION

The Royal Yachting Association offers appropriate levels of experience for Supervisors. The Supervisor must be familiar with the contents of Chapter 6.2 - Supervision and have the approval of the Operating Authority.

The RYA Dinghy or Keelboat Instructor is an appropriate qualification, or the equivalent level of experience for ventures on inland waters. The Supervisor must have considerable experience of water similar to that being used and be competent in assessing water and weather conditions. Sailing dinghy ventures which take place on rivers, canals and navigations can usually be supervised from the bank providing the Supervisor has road transport and is able to meet the group at suitable locations during the day and at camp sites.

Where ventures take place in sheltered estuaries, the RYA Advanced Dinghy Instructor or RYA/DTp Yachtmaster Offshore Certificate of Competence are the appropriate qualification, or an equivalent level of experience. The Supervisor must accompany the venture in a safety boat. Contact should not be made with the group during the journey except for the needs of supervision. The safety boat must be sufficiently remote from the participants to avoid destroying the group's sense of remoteness and self-reliance and yet be able to render assistance in an emergency within a reasonable amount of time.

Written approval should be obtained from the Operating Authority for the venture and its supervision.

ASSESSMENT

The Assessor must be familiar with the contents of Chapter 6.4 - Assessing Qualifying Ventures, briefed for the role, and should be accredited if possible. Assessors should have the same qualifications or equivalent experience as Supervisors and be familiar with the water being used for the venture.

VENTURES IN YACHTS

An Expedition in a yacht is a serious undertaking and may be beyond the resources of most young people within the Award Scheme, unless one of their parents has a suitable boat or one is accessible through the Operating Authority. Department of Transport regulations concerning sail training yachts have been tightened and the need for ancillary qualifications, such as VHF Radio Operator, all adds to the challenge. Like Other Adventurous Projects, yacht ventures are best confined to those aged 18 and over and who are not in *loco parentis*. It is acknowledged that many of our finest competitive sailors are in the 18 to 25 age range of Award participants. To be able to handle a yacht competently is a most satisfying and rewarding experience and, for many sailors, represents the fulfilment of their aspirations.

REQUIREMENTS

The requirements for yacht ventures are identical to those for all other ventures except that the participants will normally sleep aboard the vessel. See Chapter 2.1 - Introduction and Requirements.

Because these ventures usually take place in open estuaries, sea lochs, offshore or in open sea, the Award Scheme requires the Supervisor and/or the Assessor to be aboard and hold the RYA/DTp Yachtmaster Offshore Certificate but the vessel must be sailed, helmed, navigated and skippered by the participants themselves, each taking on all the various duties in turn. As with ventures on land, not all members of the crew have to be under assessment.

The requirements of the Expeditions Section specifically exclude the use of all motorised assistance during the venture. The one exception is the use of an auxiliary engine, or outboard motor, on a yacht when custom, practice and good seamanship dictates that an auxiliary motor should be used. **Yachts may not be used at Bronze level.**

CONDITIONS
SILVER

Duration: Twenty one hours of planned activity over the three days, averaging seven hours a day. The three days may be part of a longer period of sailing.

Water: Inland waters, estuaries or sheltered coastal waters which must present an appropriate challenge and are **unfamiliar** to the participants.

Competence: All participants must attain the RYA Day Skipper Shore Based and Practical Certificates or have an equivalent level of competence.

GOLD

Duration: Thirty two hours of planned activity over the four days, averaging eight hours each day. The four days may be part of a longer passage or period of sailing.

Water: Inland waters, lochs, estuaries or sheltered coastal waters which must present an appropriate challenge and are **unfamiliar** to the participants. Open sea areas may be used by suitably experienced crews.

Competence: All participants must attain the RYA Day Skipper Shore Based and Practical Certificates or have an equivalent level of competence.
For offshore ventures it is mandatory to hold the appropriate award.

For ventures in vessels making overnight passages in open sea areas, at least one crew member must hold the RYA/DTp Coastal Skipper Certificate of Competence.

EQUIPMENT

Yachts must be in a sound condition and equipped to modern good standards of custom and practice, with suitable life-saving and emergency equipment. They should comply with the appropriate current RYA recommendations for the type of boat.

A venture in a yacht is a serious undertaking
J. Driscoll

Personal clothing and equipment may be based on the appropriate lists for sailing dinghies.

TRAINING

The Common Training Syllabus as set out in the *Award Handbook* should be followed in conjunction with the advice given in Chapter 4.4. - Water Ventures - An Overview. Marine navigation should be substituted for land navigation. The training related to sailing skills should be at the levels listed above. Yachts usually accommodate the participants, so the camp craft syllabus should be modified to attain high standards of galley cooking and a personal routine which will keep the vessel tidy and shipshape.

SUPERVISION

The Supervisor must be familiar with the contents of Chapter 6.2 - Supervision and be approved by the Operating Authority.

Where ventures take place in inland waters, lakes, lochs and sheltered estuaries, the Supervisor should, where safe and practical, accompany the venture in another vessel. Contact should not be made with the group during the journey except for the needs of supervision. The Supervisor must be sufficiently remote from the participants to avoid destroying the group's sense of remoteness and self-reliance and yet be able to render assistance in an emergency within a reasonable amount of time.

For ventures on sheltered coastal waters and open sea areas the Supervisor should be aboard the vessel and hold the RYA/DTp Yachtmasters Offshore Certificate of Competence but should not intervene unless for reasons of safety.

Written approval should be obtained from the Operating Authority for the venture and its supervision.

ASSESSMENT

The Assessor must be familiar with the contents of Chapter 6.4 - Assessing Qualifying Ventures, briefed for the role, and should be accredited if possible. Assessors should have the same qualifications or equivalent experience as Supervisors and be familiar with the water being used for the venture.

For ventures on sheltered coastal waters and open sea areas, the Assessor should be aboard the vessel and hold the RYA/DTp Yachtmaster Offshore Certificate of Competence where possible. Providing Assessors have extensive yachting or keelboat sailing experience they may also fulfil the role of Supervisor provided they have the approval of the Operating Authority.

Where Assessors hold the necessary qualifications, and are approved by the Operating Authority, they may also take on the role of Supervisor and should be totally independent of the Award Unit at Silver level. At Gold level they **must** be totally independent of the Award Unit and not associated with it in any way. Where a person takes on the dual role of Supervisor and Assessor it halves the amount of adult intervention in the journey and enhances the participants' feeling of self-reliance.

Rowing

This Chapter must be read in conjunction with
Chapter 4.4 - Water Ventures - An Overview.

Rowing ventures represent a satisfying and safe means of travelling on water. The level of skills required, though challenging, can be achieved within the time span of each of the levels of the Award. A rowing venture in a gig or whaler must epitomise the concept of teamwork where a group has to work in harmony to ensure their mutual success. The fact that very few groups undertake rowing Expeditions or Explorations is most probably due to the difficulty in finding suitable boats in which to undertake the venture and in their transportation.

Rowing ventures must take place in a boat which has been designed primarily for rowing. The relatively broad beam of sailing dinghies, their design and lightness makes them extremely frustrating and tiresome to row over an extended period of time such as on an Award Expedition. There must be many thousands of rowing boats of all shapes and sizes ranging from those capable of carrying pairs of participants and their gear, to gigs, whalers and ex-ships' lifeboats which can carry a group of seven on rivers, canals and sheltered estuaries. It would surely repay the enterprise of Award Units who have the opportunity to acquire such boats to restore and bring them into use for their own groups; they might even be able to recoup some of their expenses by hiring them out to Award groups who are less fortunate than themselves.

The water suitable for rowing ventures is best compared with that used for canoeing within the Expeditions Section and includes rivers, canals, other waterways, navigations and sheltered estuaries. There are endless opportunities on the Continent without many of the problems of access associated with British rivers - large pulling boats have already been trailed across the Continent to take advantage of these facilities. The sheer size and magnificence of the larger continental rivers such as the Loire, Danube and Rhône make them an inspiring challenge to anyone in a small boat.

Though it may be helpful to compare canoeing water with rowing water within the Expeditions Section, it must be remembered that rowing boats cannot be man-handled for any distance and it is essential that there are locks at all weirs and sluices in addition to slipways for launching and recovery. The water used for rowing ventures must be unfamiliar to the participants at Silver and Gold levels but this does not preclude reconnoitring beforehand from the bank. It is essential that access to the water, launching and recovery points, mooring sites and camp sites be identified before the venture. Using a lock may take up to 20-30 minutes and, if there is a lock-keeper, use may be restricted to the working day, frequently with a break for lunch; all this must be determined beforehand and built into the planning. Pulling boats require the use of locks. It is essential to seek written permission to use locks if a venture is to take place on canals on the Continent or where the rivers has been canalised or turned into navigations. This should take place at the same time as access to the water and camp sites is being sought. The relevant National Tourist Office should always be consulted before taking boats to the Continent.

REQUIREMENTS

All the requirements for rowing ventures are identical to those for all other ventures. See Chapter 2.1 - Introduction and Requirements.

CONDITIONS

The group must consist of between 4 and 7 participants but there are no restrictions on the number of boats which may be used.

BRONZE

Duration: A minimum of four hours of rowing time on each of the two days.

Water: Rivers, canals and other inland waters in rural areas.

There is no distance requirement as such, but a reasonable expectation would be for a group to cover 16-20 kilometres (10-12.5 miles) on each day depending on factors such as the number of locks, head winds and the strength of the current.

SILVER

Duration: A minimum of five hours of rowing time on each of the three days.

Water: Rivers, canals and other inland waterways and lakes in rural areas. The water must be **unfamiliar** to the participants and present an appropriate challenge.

There is an expectation that the conditions will be related to the age and experience of the participants and represent a progression between Bronze and Gold. The five hours' rowing each day would realistically cover 65 kilometres (40 miles) over the three days.

GOLD

Duration: A minimum of six hours of rowing time on each of the four days.

Water: Rivers and navigations in rural areas and sheltered coastal waters and estuaries. The water must be **unfamiliar** to the participants and must present an appropriate challenge.

At Gold level there is no expectation that the water should be more difficult than our major navigable rivers. Moving water, either by current or tide, should be sought where possible. Six hours of rowing each day would realistically cover 32 kilometres (20 miles) according to the current or tide, resulting in a journey of around 128 kilometres (80 miles) in all.

TRAINING

The appropriate level of the Common Training Syllabus, as set out in the *Handbook* should be used in conjunction with the advice given in Chapter 4.4 - Water Ventures - An Overview. For rowing ventures on rivers, canals and navigations, the standard land navigation syllabus will suffice, but for sheltered estuaries appropriate changes should be made.

There is no relevant training scheme or system of awards for this form of rowing similar to those provided by Governing Bodies such as the British Canoe Union or the Royal Yachting Association, but the Sea Cadets, Scouts, Guides and some Local Authorities have effective training schemes of their own which provide a suitable basis of training for Award groups operating within these organisations. The Boatwork programme of the Skills Section of the Award Scheme has provided the basis of the training programme in the past, but a more detailed syllabus has now been prepared and is set out below. Parts A and B are sufficient for ventures on canals, rivers and navigations but Part C is mandatory for all ventures on tidal waters. All participants should be of the same level of competence whatever the type of rowing boat involved, but two or three pairs of friends in their small rowing boats will be more informal than the more organised structure demanded from six or seven participants all in the same gig or whaler.

BOATWORK SYLLABUS
PART A
All participants must:
- Be able to swim 25 metres without any buoyancy aid.
- Wear a life-jacket or buoyancy aid when on or near the water.*
- Use a boat fitted with adequate buoyancy which is firmly secured.
- Be able to tie the following knots:
  clove hitch, round turn and two half hitches, bowline, reef knot, figure of eight and a quick release knot. It is far more important to be able to tie half a dozen knots well, than to know ten or twenty and not be sure which one to use or how to tie them. Participants must be able to tie knots quickly, without thinking, without looking and from any position.
- Be able to make fast to cleats, posts, pins and rings and to anchor.
- Be able to throw a line accurately and reliably.
- Be competent in man overboard drill.
- Know how to use locks, if appropriate for the venture.
- Be able to carry out maintenance and simple emergency repairs.
- Ensure that the boat is properly equipped, including the minimum of essential spares.
- Be able to stow the gear and trim the boat.
- Waterproof food and clothing and other equipment which may suffer from immersion or rain.
- Be able to fulfil the role of cox, stroke, bow and crew member as appropriate.
- Be able to give and respond to orders in a seaman-like fashion to work together to handle a boat under oars - getting away from quay, jetty, bank, moorings, and beach; pulling and backing; coming

Rowing boats are too heavy to be manhandled for any distance

alongside; picking up moorings; manoeuvring the boat without use of the rudder; recovering man overboard and beaching; handling a boat under tow; taking a boat in tow, and allowing for wind and current when carrying out these manoeuvres.

- Be able to carry out a ferry-glide where pulling boats are to be used on rivers with strong currents or estuaries with strong tidal flows (2-3 knots or more).

*Note: It is tradition, custom and practice not to wear lifejackets in pulling boats, except under potentially dangerous conditions at the discretion of the coxswain, because they impede movement and can be very hot and uncomfortable in warm weather. The Sea Cadets, Scouts and Guides do not wear them as a rule unless ordered by the coxswain. All evidence indicates that this is a reasonable and safe practice. **Participants must obey the requirements of their Operating Authority**. Those who are only concerned with rowing for the preparation and execution of a particular level of Award are strongly advised to use modern buoyancy aids, similar to those used by canoeists, at all times. They do not restrict movement to the same extent as traditional life-jackets and are not as uncomfortable.

For a pulling boat, there must be a series of commands such as: 'back oars', 'hold water', 'starboard hold water', 'give way together', 'ship oars' and 'bow oar' so that the boat can be manoeuvred efficiently to operate safely and in a seaman-like manner. A comprehensive list of Boat Orders may be obtained from the Scouts, Guides and Sea Cadets for a small charge. During training and the qualifying venture, there should be a regular rotation of positions and roles to develop an all-round competence and to provide a refreshing change of task.

PART B
In addition to Part A all participants must know:
- The proper names and function of all the parts of a pulling boat.
- The basic boating, sailing and sea terms.
- The basic rules of the road and priorities.
- The sound and distress signals used on water.
- The courtesies, customs and etiquette associated with boating and sailing.

PART C - TIDAL
In addition to Parts A and B, participants who carry out their venture in tidal estuaries and sheltered coastal waters must be competent in the following:
- Navigation, using the relevant charts and navigational information.
- Tidal processes, ebbs and flows, streams, rise and fall, range, springs and neaps.
- The use of publications providing tidal information.
- The relevant buoyage system.
- Handling a boat competently in tidal waters, including mooring and beaching.

The syllabus outlined above should form the basis of the training programme for rowing ventures when used in conjunction with Chapter 4.4 - Water Ventures - An Overview.

EQUIPMENT
The choice of equipment will be determined to some extent by the water being used but the following list will serve as the base.

General Equipment Related to the Boat
- All boats should be in a sound condition and properly fitted out with stretchers and long painters, fore and aft, for use in locks.

- Crutches must be tied to the boat and a spare oar and crutch should be carried.
- Bucket, bailer and sponge.
- Throwing line.
- Spare line.
- Charts/maps - in water-resistant protection.
- Compass(es).
- Repair kit including canoe tape, very thick flexible plastic sheeting etc.

If the venture is to take place in a sheltered estuary then compass(es), distress flares (red signal and smoke), water and emergency rations should be carried and a powerful water-resistant torch, air horn and a buoyant knife may be useful.

Personal and Expedition Equipment

Participants should refer to the equipment list in Chapter 3.1. Most equipment is the same for all modes of travel with adjustments being made where necessary.

Clothing

This list may be used as a checklist with the necessary substitutions, but participants should remember that it can become very cold around the camp site at night:
- Suitable clothing including protection for the head, hands and feet.
- A complete change of clothing in a waterproof container.
- Protection for the feet which must be worn at all times; deck shoes, plimsolls, plastic sandals or even old trainers are ideal.

Personal and Emergency Equipment

This list has been amended to make it more relevant to the sailors' needs:
- Lifejacket or buoyancy aid with attached whistle.
- Bivvy bag (poly-bag).
- Waterproof or water-resistant watch.
- Sailor's knife with blade, tin-opener and spike.
- Matches in waterproof container.
- Personal first aid kit.
- Emergency rations.
- Small water-resistant pocket torch with spare bulb and alkaline batteries.
- Notebook and pencil.
- Coins/phonecard for the telephone.

Clothing may have to range from waterproof over-clothing to protection from the sun. If the weather is cold then the extremities, head, feet and hands, must be insulated and remember that coxing can be a cold task. Gloves may help to prevent blistering and some may find padding on the seat beneficial.

WATERPROOFING EQUIPMENT

Clothing, sleeping bags and food should be waterproofed in drums or dry bags, in a similar fashion to that used on a canoeing venture, so that they are not affected by immersion or rain. It will also help to keep the boat tidy and uncluttered. Snacks and drinks in a plastic container should be readily available to the crew while rowing. When packing, crews should bear in mind that they may have to carry their equipment from the mooring to the camp site. The size and weight of the packs must be considered carefully.

SUPERVISION

The Supervisor must be familiar with the contents of Chapter 6.2 - Supervising Qualifying Ventures and be approved by the Operating Authority.

There is no recommended qualification or equivalent levels of competence for Supervisors except for the Sea Cadets, Scouts and Guides who have their own certifcate systems. Those not involved with these organisations might look to them for appropriate levels of competence, or to The British Canoe Union or The Royal Yachting Association for equivalent levels of experience. The Supervisor must have considerable experience on the type of water being used and be competent in assessing water and weather conditions. Rowing ventures on rivers, canals and navigations, as with canoeing ventures, can be readily supervised from the bank, providing the Supervisor has road transport and can meet the group at locks or bridges during the journey and at the camp sites in the evening.

Written approval should be obtained from the Operating Authority for the venture and its supervision.

Where ventures take place in sheltered estuaries, the Supervisor must accompany the venture in a safety boat. There should be no contact with the group during the journey except for the needs of supervision. The safety boat must be sufficiently remote from the participants to avoid destroying the group's sense of remoteness and self-reliance and yet be able to render assistance in an emergency within a reasonable amount of time.

ASSESSMENT

The Assessor must be familiar with the contents of Chapter 6.4 - Assessment, briefed for the role, and should be accredited if possible. The assessment of rowing ventures on rivers and canals should not present any difficulty and Assessors able to assess canoeing or sailing ventures should be able to cope with rowing ventures. For ventures in sheltered estuaries the Assessor should be experienced in the chosen mode of travel and familiar with the water involved in the venture.

EXPEDITION GUIDE

5

THE DUKE OF EDINBURGH'S AWARD

YOUR

VENTURE –

PREPARATION

AND

PLANNING

Neil Campbell-Sharp / National Trust

Part 5

Your Venture - Planning and Preparation

Young People with Disabilities

The personal challenge and non-competitive nature of the Award programme, where young people are assessed on their own progress, perseverance and achievement, makes the Scheme very attractive to young people with disabilities.

There is one Award with a common philosophy, aims and objectives, a universal structure, timescales and age requirements which must be observed by all participants. Flexibility within the Scheme enables young people with disabilities to take part on a comparable basis with more able participants and to reap the benefits.

The *Award Handbook* states that all who choose to participate in the Scheme should be capable of meeting the challenge and should:

- Make a commitment and have a basic ability to communicate.
- Have sufficient knowledge and skills, self-reliance and self-orientation, particularly when meeting the requirements of the Expeditions Section.
- Choose their own activities.
- Have the will to take part in the Scheme and not just a willingness to be led through it.

The full requirements and conditions are given in the Expeditions Section of the *Award Handbook*. The requirements are the demands which apply to all ventures at all levels, while the conditions relate to specific obligations which have to be fulfilled at the three levels of the Award - Bronze, Silver or Gold.

These requirements and conditions may be summarised as:

- Participants must complete an unaccompanied self-reliant journey of the required distance or hours of travelling time, by their own physical effort, without motorised assistance in the prescribed number of days, and then submit a presentation of their venture. While the ventures must not be accompanied they must be supervised by a suitable person to ensure that the high standards of safety are maintained.

The requirements and conditions ensure that the value and integrity of the Award is maintained for past, present and future Award holders. It is essential that both the spirit of the Award, as well as the conditions, are observed by all participants.

To many young people with disabilities, the requirements and conditions may appear daunting, but this is not necessarily so as there is sufficient flexibility within the requirements to accommodate the majority of young people. All should, wherever possible, endeavour to meet the requirements and conditions of the Award, relying on the flexibility within the Expeditions Section to accommodate their needs. This may involve considerable extra time and effort on the part of the participants, the Instructors and the Supervisors, but the sense of achievement and the knowledge that they have faced the challenge on similar terms with all the other participants will be their reward.

USING THE FLEXIBILITY OF THE EXPEDITIONS SECTION
Distance

For many young people with disabilities the completion of the minimum distance for land ventures may present the greatest obstacle. There are alternatives.

Ventures on water by canoe, rowing boat, sailing dinghy or keelboat, as well as ventures on horseback, do not involve a mandatory distance, the challenge being determined by a prescribed number of hours of travelling each day. For example, at Bronze level, canoeing, rowing and horse riding all involve travelling for 4 hours a day while for sailing

The careful selection of routes is the key to successful ventures

ational Trust

ventures there should be 12 hours of planned activity over the two days. Increasing use is being made of these exciting forms of travel by young people with disabilities, especially by those with an impairment of the function of the lower limbs. The use of open canoes where - 'if you fall in, you fall out' - on canals, placid or Grade 1 water is a common form of venture in the Expeditions Section and is usually readily available.

There are few, if any, restrictions on the type of terrain at Bronze and Silver levels other than that ventures must take place in normal rural or open country. Rural country offers an endless variety of paths, bridleways and tracks with a realistic choice of gradients which may be negotiated by wheelchair. In addition, there are many hundreds of miles of towpaths along canals which are virtually free of any gradient while meandering through rural countryside to connect our major cities.

At Gold level, Expeditions must take place in one of the Award Scheme's designated Wild Country Areas, but even here there are large tracts of level terrain in the valley floors and on the moorlands or plateaux at higher altitudes. Parts of Wild Country Areas are often have with many miles of tracks. In addition, in some of these areas there are canals with their accompanying towpaths, as well as disused railway lines which have been turned into bridleways or cycle tracks. The careful selection of a route is probably the most crucial factor in ensuring the success of ventures for those with disabilities.

Explorations may provide an alternative option, where the travelling time may be reduced to a minimum of 5 hours at Bronze and 10 hours at Silver and Gold levels during the venture. This will provide more time for study, observation and recording associated with the purpose of the venture, as well as an increased opportunity for the awareness of nature and the environment in which the journey takes place.

Load Carrying

Where load carrying presents a problem, the preferred option should be to choose a mode of travel such as canoeing, rowing or sailing. If this is not possible then, with the written permission from the highest level within their Operating Authority, and not at Unit level, certain individuals within a group may have some equipment and food pre-positioned at camp sites. Listed emergency equipment, individual sleeping bags and some food must always be carried for safety.

Accommodation in the Expeditions Section is by camping, but again, with the written permission of the Operating Authority, basic accommodation such as barns, bothies or mountain huts may be allowed. This could result in a considerable reduction in the weight of the load to be carried. Operating Authorities may find it useful to discuss any plans to allow accommodation other than by camping with their local Territorial or Regional Officer. Explorations at Silver and Gold levels allow the same camp site to be used on more than one night which reduces the load carrying over the venture as a whole.

Learning Difficulties

It is expected that participants with learning difficulties will complete a journey of the usual distance or hours of travelling, carrying a standard load.

Some young people with learning difficulties may need more instruction over a longer period of time to become confident in navigation, camp craft and safety and emergency procedures. For land-based ventures,

Alan Lewis

most travelling takes place on footpaths, bridleways and tracks. Again, a careful selection of routes which are easy to follow is usually the solution. Paths alongside rivers, canal towpaths, and disused railway lines where there is a right-of-way are obvious options. Simplified maps or itineraries may be prepared to supplement Ordnance Survey maps.

Integration

Groups must be of between four and seven young people in the Expeditions Section. While there is an expectation that participants will be of roughly the same age and experience, the integration of those with disabilities into a group intending to fulfil the standard conditions has much to commend it, particularly if it is an existing friendship group. It enables many with learning difficulties to play a fully integrated part in the venture, complying with the standard conditions.

Integration must always take place at the earliest stage in the preparation of any venture. This *Expedition Guide* sets great store in young people taking possession of their own ventures and forming their own groups. It would be wrong for a group to be pressurised into accepting another member at the last minute.

It is essential that all are challenged at a level relating to their personal capabilities. Participants with mobility difficulties may be totally integrated into horse riding, cycling or water ventures. Those with disabilities need to be trained to their full potential and each individual should be able to make a significant contribution to the venture in a diversity of ways. Assessors should be notified that the group includes participants with disabilities before the commencement of the venture and, for ventures at Gold level in wild country, this should happen when the green *Expedition Notification Form* is dispatched to the Wild Country Panel Secretary.

Supervision

Groups must not be accompanied by any adults but they must be supervised by an appropriately experienced person. To ensure opportunities for independence and self-reliance, visits should be limited to those which are really essential. There is a considerable latitude in the number of times a Supervisor may make contact with the group while en route or at the camp site each day. Supervisors may observe a group from a distance and visit the group as often as necessary to ensure the safety and well being of the participants.

AUTHORISED VARIATION FOR PARTICIPANTS WITH PHYSICAL AND SENSORY DISABILITIES

While the majority of participants can be accommodated by the flexibility and integration which exists within the Award Scheme, there are some who may still be unable to fulfil particular requirements set out in the *Award Handbook*. For participants with a permanent physical or sensory disability it is possible to make an application to the Operating Authority to vary specific conditions which cannot be met by any of the usual methods. Where appropriate, this application should be made in consultation with the participant's medical adviser, and parents/guardians must be involved at all times for those under 18 years of age. When considering variations it is beneficial to seek the advice of the appropriate Award Secretary or Regional Officer.

At Bronze and Silver levels, if it is impossible to complete the mandatory requirements, a variation may be approved by the Operating Authority, providing there is some compensation such as extra time spent on the purpose work of the venture.

At Gold level, the Award Secretary/Regional Office or Award Head Office must be involved in the consultative process to ensure that participants receive effective guidance based on examples of good practice. The overwhelming need must be to ensure that the health, welfare, safety and well-being of the young person is not jeopardised in any way.

GENERAL CONSIDERATIONS

There are certain considerations which must be observed in all ventures involving those with disabilities. For integrated ventures, it is essential that the challenge faced by the rest of the group is not diminished. A venture at Gold level using a route suitable for a wheelchair may not be sufficiently physically demanding for able-bodied participants. Alternatively, it may be possible to form a group of young people who are all in wheelchairs. There have been many good examples of this practice.

Young people with disabilities should always be considered on an individual basis. Each individual has a unique set of talents and abilities to contribute to the team effort. The venture should always be centred around their abilities and not around their disabilities.

Parents should be kept informed of the details of the venture and of the nature of the supervision. Their advice and involvement should always be sought.

Many participants with disabilities may well have very little experience of the outdoors, and even less confidence. Confidence-building is an essential preparation before training can begin. Care and sensitivity is needed to ensure that positive feelings towards the outdoors are engendered and that the subsequent journeys are going to be enjoyable and rewarding experiences. Youth Hostels and frame tents may provide a suitable introduction to camping and, for some, a frame tent may be essential during the qualifying venture.

Any special items of equipment required to support a participant must be checked with the same care as all Expedition equipment to ensure that it is in a sound condition, is suitable for the purpose of the venture and sufficiently robust to stand up to the task involved.

In Chapter 6.1 - Training, the importance is stressed of the repetition of certain skills until they become drills which bring about an automatic response to a threatening situation. It may be necessary to use the same technique to deal with certain aspects of camp routine to ensure the safety of individual participants with disabilities.

As mentioned previously, the choice of route is of critical importance. The same attention must be devoted to communications and the choice of camp site, where ease of access and other facilities may be of considerable importance.

The preparation and submission of the presentation should not be a problem providing it has received attention at an early stage of training. There is so much choice for this process that all young people should be able to produce an interesting and creative presentation of their experience.

Unit Leaders, Instructors and Supervisors concerned with young people with disabilities should endeavour to liaise with other Award Units who are faced with similar situations so that experience can be shared. This will usually be through the Operating Authority but, if this is not possible, then the help of Award Secretary/Regional Officer should be sought. They may be able to advise Leaders and Instructors where to establish contact with other groups involved with young people with disabilities.

Adults involved with participants who have disabilities have a challenging task and one that requires an involvement over a long period of time. This requires a great deal of dedication, counselling skills and patience. There are considerable rewards in extending the breadth and scope of the young people's experience, their sense of fulfilment and feelings of achievement which may well alter their whole perception of their strengths and abilities in later life.

David Boag: Oxford Scientific Films

William Fediw

Effective Practice Journeys

Many Award groups fail to obtain the full benefit from practice journeys, which represent, for many, the greatest missed opportunities of the Expeditions Section. The serious problems which sometimes arise in the qualifying ventures could be avoided if more effective use was made of the practice journeys. Problems associated with poor time-keeping, overweight packs, lack of physical fitness and blistered feet could all be reduced or eliminated if the relevant lessons had been learnt during training. Practice journeys should be enjoyable and rewarding for, in terms of experience, skill, and personal and group development, they are of equal importance to the qualifying venture and should always be considered as having great merit in themselves rather than being just a means to an end.

Following initial training all participants are required to undertake sufficient training and practice journeys to ensure that they have acquired a level of competence to be able to safely complete an unaccompanied, self-reliant venture.

In order that the Award Leader or instructor has the opportunity to judge the participant's experience and competence, and to give the group the opportunity to work together as a team, **at least one practice journey must be undertaken** at each level of Award.

This practice journey must not be over the same route or in the same vicinity of the route to be used during the qualifying venture, but the conditions should be as similar as possible to those anticipated, including the distance each day, and should be undertaken in terrain which is equally demanding. Practice journeys at Silver and Gold levels should include two or more nights camping. **For qualifying ventures in wild country this practice journey must be in a wild country environment.**

Practice Journeys must be the same mode of travel as the qualifying venture.

All participants must have an opportunity to experience unaccompanied journeying before undertaking the qualifying venture. Unaccompanied practice journeys must be supervised and, if in wild country, the appropriate Wild Country Panel informed using the standard *Expedition Notification Form* available from Operating Authorities.

It is recommended that:

• The final practice journey should take place in the same expedition season as the qualifying venture.

• Practice journeys are not undertaken immediately prior to the qualifying venture as this makes unreasonable demands on the participants and does not allow time to reflect upon or initiate any changes required to the planning.

It is inappropriate to prescribe the number of other practice journeys to be undertaken in order to reach the level of competence required. The final practice journey should be seen as the culmination of a period where the group has worked together in planning and preparation, in undertaking accompanied practice journeys with the Leader or instructor before embarking on unaccompanied ventures with supervision. The Leader's judgement is crucial in this respect and there are no short cuts where the safety and well-being of the participants is concerned.

There are great advantages in changing the mode of travel at different levels of the Award insofar as it increases the excitement and challenge, extends the breadth and depth of experience and presents new skills to the participants. This would usually necessitate one or two additional practice journeys - a price well worth paying!

It is desirable that practice journeys are of the same duration as the qualifying venture, but this is not mandatory. There is great merit in having two nights' camping at Silver and Gold levels, though one will suffice at Bronze level. Anyone can camp for one night; lessons are learnt from coping with wet camping gear and clothing, carrying the wet gear and then using it the following night.

To derive the most effective benefit from practice journeys, it is essential that clear aims and objectives are established beforehand for each journey. Participants and Instructors must have a clear expectation of the learning outcomes. Practice journeys provide the ideal opportunity for skills and techniques which have been taught in isolation to be integrated into a total experience and practised under realistic conditions. Practice journeys are an integral part of the team building process as well as providing facilities for the reviewing process. In addition to the general aims, it is essential that all involved should have clearly defined training objectives.

Groups should have a preparation session before every practice journey where the purpose of the journey and specific training needs are clearly identified.

There is great merit in camping for two nights on practice journeys

Joe Cornish

Participants may wish to use practice journeys to:
- Practice their expeditioning skills with particular emphasis on the mode of travel, navigation and camp craft.
- Determine their speed of travel in a similar environment and under the same conditions as their qualifying venture.
- Test the suitability of their clothing, footwear and equipment and the weight of their packs, and eliminate any unnecessary items of equipment.
- Find out how much food, drink and fuel they will need to cope with the expenditure of energy during their venture.
- Appreciate the levels of fitness required for the qualifying venture and the physical endeavour involved.
- Practise skills of observation and recording related to the purpose of their venture which they will use in the report on their qualifying venture.
- Develop the interpersonal relationships and team building processes which are essential to survive a period of sustained physical endeavour in the face of possible foul weather in a demanding environment.

The Instructor may wish to use practice journeys to:
- Provide individual and group tuition in expedition and travelling techniques over a number of days where the participants have an opportunity to establish an apprentice relationship with the Instructor.
- Evaluate the competence of the group and the individuals to undertake an unaccompanied, self-reliant venture.
- Observe, and facilitate where necessary, the team building process.

THE TIMING OF PRACTICE JOURNEYS

Second only in importance to clear aims and objectives is the timing of practice journeys. Training for qualifying ventures takes many forms; frequently it consists of weekly sessions from the Autumn through to the following Summer.

Where practice journeys take place too early, participants have too few skills to derive the greatest benefit. When they take place shortly before the qualifying venture there is insufficient time for any lessons to be learnt and shortcomings in equipment, training and fitness to be rectified. **Journeys which take place immediately before the qualifying venture may well jeopardise an individual participant's chance of**

success, or even that of the whole group. While such timing is not expressly forbidden, it may result in participants embarking on their qualifying venture feeling exhausted, with badly blistered feet, wet equipment and inadequate sleep. This practice is usually designed to save time, travel and money, but it should be strongly discouraged, especially at Gold level in wild country. **Where practice journeys take place immediately prior to the qualifying venture, the consequences for all the participants must be fully considered beforehand and adequate time allowed for recovery.**

Practice journeys may take place over a weekend for Gold level ventures, though a long weekend is more appropriate. It is essential that travelling distances are sufficient to present a realistic physical challenge to the participants. It may be possible to travel to a camp site on a Friday evening, set up camp and prepare a meal; if it has to be done in the dark, so much the better. This can be followed by a full day of travelling and camping for a second night. A very early start on the Sunday morning, a journey lasting to the middle of the afternoon and a return to base will provide a demanding practice. The journey should always be followed by a reviewing session, where the experience is consolidated forming a base for future planning and training.

GENERAL CONSIDERATIONS

Practice journeys represent peak experiences in the training programme. Participants need these journeys to find out vital things such as whether their speed of travel and physical fitness on terrain, or water, is up to that required on their qualifying venture. They need to know how much food they will require each day, the suitability of their clothing and equipment, and their ability to navigate.

Team building is enormously enhanced where members of a group work together over a period of time. Practice journeys, like qualifying ventures, provide excellent opportunities for the development of interpersonal relationships as well as time for the participants to develop leadership skills.

Practice journeys enable participants to practise observing and recording skills and to evaluate their suitability for their qualifying venture.

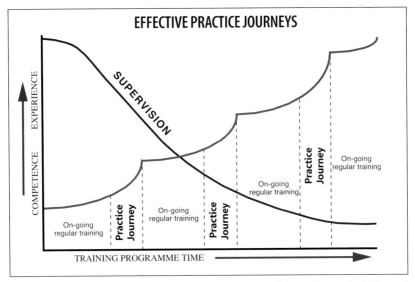

Instructors have their best opportunity to evaluate the suitability of equipment, the participants' competence in realistic conditions and their ability to cope with a self-reliant journey. This enables them to sign the *Record Books* with confidence.

Nowhere within the Expeditions Section is the process of reviewing more important. At natural breaks during the journey and in the evening at the camp site, there will be ample opportunities to review the activity and see how it is meeting the needs identified before the journey. Discussion may lead to more effective strategies which may be implemented during the venture and reviewed in their turn, or may identify the need for further training before the qualifying venture. Of equal importance is the need to review the performance of individuals, teamwork and the functioning of leadership within the group.

Groups should meet together independently of their Instructor and review their needs. A list of needs should be prepared by both the group and individuals within the group. They should decide how these expectations can be met and devise training to accomplish their objectives.

Extra practice journeys may always be added to the training programme, but the mandatory practice journeys, which have to be entered and signed in the *Record Book*, must always conform as closely as possible

Tony French (Halina/Fuji Bursary)

Practice journeys are essential for determining the speed of travel

with the qualifying venture, with full equipment being carried, to ensure the safety of the participants.

Carrying out more than the minimum number of practice journeys enables Instructors to reduce the amount of direct supervision as the group progressively becomes more self-reliant until they feel confident that the group can travel unaccompanied, dependent on their own abilities and resources, which is one of the principal objectives of the Expeditions Section. There is great merit in Instructors accompanying the first practice journey at Bronze, Silver and Gold levels, either throughout, or for a considerable part of the time, to enable them to evaluate the participants' needs and monitor their progress.

During an accompanied venture, the Instructor will be able to delegate route finding and other tasks to individuals according to their needs, insisting on each individual being able to locate their position on the map, track progress along the route, set the map at all times and identify places when requested. There are endless opportunities for remedial teaching of the skills common to all ventures as well as those related to the mode of travel.

Where Instructors accompany practice journeys for all or part of the journey, they have the opportunity to facilitate and monitor the team building process and assign the leadership role to individuals in turn, as well as being able to participate in the reviewing process. Instructors must bear in mind that the final practice journey should be unaccompanied to allow the participants the opportunity to work together as an independent team before their qualifying venture.

Instructors may wish to refer to Chapter 6.1 - Training which will assist them in maintaining high levels of safety as participants are prepared for unaccompanied self-reliant ventures.

Practice journeys for Explorations and Other Adventurous Projects should conform to normal practice and fulfil the conditions listed above. For Explorations the minimum journeying time may be reduced if this is appropriate for the nature of the venture; this will make it possible to devote more time to the purpose.

Planning Your Qualifying Venture

All ventures in the Expeditions Section belong to the participants and, ideally, all should be able to complete a venture of their own choice. A significant difference between this Section of the Award and the others is that the Expedition is a group effort and the success of the individual participants is dependent on the success of the other members of the group. Participants must work together to plan a joint venture which will meet the aspirations of the individuals within the group.

The choice of companions is important for it may well determine whether the qualifying venture is a success. The process of forming a group may be based initially on personal friendships, and a desire to fulfil a common purpose by a mutually preferred mode of travel in a particular place. However, there are other factors which must be taken into account. Participants should seek companions who share the same enthusiasm for success and a dedication to undertake the necessary training for competence and fitness.

The intensity of experience, the ease of reaching common agreement and opportunities for leadership are always greater in a group of four, the minimum permissible. Many think that it is more prudent to be part of a larger group of five, six, or even seven, as it is not possible to predict who will fall by the wayside during the best part of a year spent on training and preparation, or who may become incapacitated during the actual qualifying venture.

The Award Scheme is a partnership of young people and Adult Leaders - Instructors, Supervisors and Assessors - working together to achieve a common purpose. The adults involved are committed and almost invariably overworked by so many young people all trying to do different things. One or two people with one minibus frequently have to cope with several groups, and a common destination is sometimes inevitable. It has been said that politics is the art of the possible; the same may be said of groups planning their Award ventures. Taking into account all the compromises which have to be made in the process of planning, every effort should be made to design a venture which reflects the aspirations of the individuals within the group. All must set out with clear ideas concerning the form and nature they would like their venture to take, even though modifications may need to be made on the way to meet the challenges of the Expeditions Section.

While groups, of necessity, may need to carry out their ventures in the same area of country as other groups from their Award Unit, there is still enormous scope for individuality. The planning of different routes which fulfil the needs of the chosen purpose, and even different modes of travel, may be possible in the same area.

PLANNING

Planning a venture usually involves answering five basic questions:

- **What?**
- **Where?**
- **How?**
- **When?**
- **With whom?**

These questions relating to the purpose of the venture, the location, the mode of travel, timing and the choice of companions are rarely made in any particular order.

Sometimes the choice may be based on a desire for the mode of travel - to carry out a venture by canoe. Other ventures will arise from the location - to undertake a venture in the foothills of the Pyrenees. Some will be designed around a specific purpose. In reality the questions are all considered together by a like-minded group of participants, with the exception of the timing of the qualifying venture, which is usually considered later as it is dependent on so many other factors.

An outline programme of preparation for the venture and its completion is shown on the following pages. The significant stages have been numbered and advice is listed below to assist in the completion of this process. This programme is intended for the most demanding ventures **taking place in wild country or on water**, but it will serve as a blueprint for the preparation of all ventures at different levels of Award, even though particular items in the programme will need to be omitted.

1. Participants should, through discussion amongst their friends and other members of the Award Unit, form a group of between 4 and 7 people who share an interest and enthusiasm for the same type of venture.

2. The group should decide on the purpose of the venture and a suitable mode of travel. It is essential that all should be available to embark on the venture at the intended time and, especially where water ventures are concerned, at the pre-planned back-up date(s).

   Opportunities exist for individuals to have different purposes during the same venture. One person may be interested in the underlying geology of the Expedition area while another may wish to record the scenic aspects of the countryside by various artistic forms of expression. Everyone in the group must share the responsibility for planning and executing the venture. Participants should make a commitment to each other to train for the venture and attain the necessary level of physical fitness. This is the time to check that suitable equipment is available for the intended mode of travel and that there are suitably experienced people available to undertake the training, as well as Supervisors and Assessors to monitor the qualifying venture when it takes place.

3. When the basic questions have been answered and the form of the venture has been established, training should commence on the skills common to all ventures, the particular skills related to the mode of travel and any training or research related to the purpose. Training usually starts in the Autumn and continues through the Winter and Spring. Where there is a change in the mode of travel, training should always commence as early as possible. For example, if there is a change from land to water, it is essential to carry out training on the water in the Autumn otherwise there may not be enough time to achieve the necessary level of competence. In any case the water around the British Isles is warmer in the Autumn than in the Spring.

AN OUTLINE PROGRAMME FOR PARTICIPANTS

Intended for ventures in Wild Country Areas but of equal significance for water ventures as well as serving as a pattern for ventures at Bronze and Silver levels

Other Adventurous Projects: submissions must reach Award Office at least 12 weeks prior to the date of departure.

Ventures Abroad: submissions on the 'blue' Expediting Notification Form, route tracings and maps must reach the appropriate Award Secretary/ Regional Office 12 weeks prior to the date of departure.

Fitness training: should begin about 3 months before departure.

Wild Country Areas: submissions on the 'green' Expedition Notification Form and route tracings must reach the Panel Secretary at least 6 weeks before departure (or 4 weeks if an Assessor is not required).

Route cards and camp site details should, wherever possible, be sent with the Notification Form, but no later than 2 weeks before departure.

Local Pre-Expedition Check: should take place in the home area 7-10 days prior to departure.

First Meeting: during the 48 hours familiarisation period in Wild Country Area prior to departure.

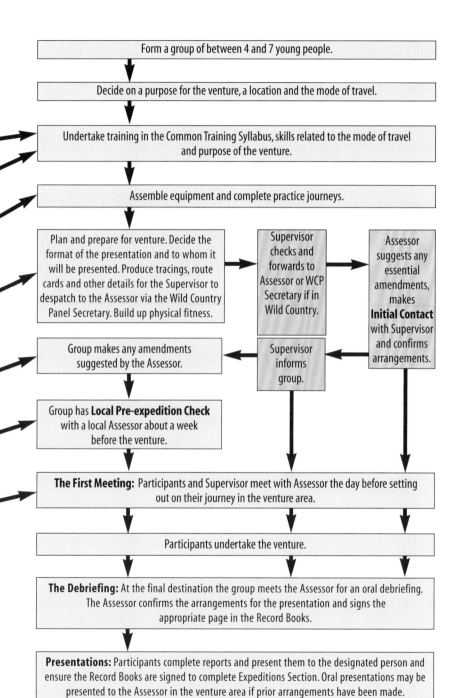

Form a group of between 4 and 7 young people.

Decide on a purpose for the venture, a location and the mode of travel.

Undertake training in the Common Training Syllabus, skills related to the mode of travel and purpose of the venture.

Assemble equipment and complete practice journeys.

Plan and prepare for venture. Decide the format of the presentation and to whom it will be presented. Produce tracings, route cards and other details for the Supervisor to despatch to the Assessor via the Wild Country Panel Secretary. Build up physical fitness.

Supervisor checks and forwards to Assessor or WCP Secretary if in Wild Country.

Assessor suggests any essential amendments, makes **Initial Contact** with Supervisor and confirms arrangements.

Group makes any amendments suggested by the Assessor.

Supervisor informs group.

Group has **Local Pre-expedition Check** with a local Assessor about a week before the venture.

The First Meeting: Participants and Supervisor meet with Assessor the day before setting out on their journey in the venture area.

Participants undertake the venture.

The Debriefing: At the final destination the group meets the Assessor for an oral debriefing. The Assessor confirms the arrangements for the presentation and signs the appropriate page in the Record Books.

Presentations: Participants complete reports and present them to the designated person and ensure the Record Books are signed to complete Expeditions Section. Oral presentations may be presented to the Assessor in the venture area if prior arrangements have been made.

Route cards and tracings for ventures in Wild Country Areas must be completed six weeks in advance

It is during this period that the interpersonal relationships and the teamwork necessary to ensure success are established.

4. Practice journeys, which usually take place in the Spring, provide the opportunity to bring together all the skills which have been acquired, frequently in isolation, to test, for instance, competence and equipment and to establish mutual confidence and trust between members of the group. Training for physical fitness should have commenced and the practice journeys will highlight any deficiencies, as well as providing the opportunity to carry out a comprehensive review of progress prior to embarking on Stage 5.

5. At this stage all arrangements will need to be finalised and intentions must be committed to paper. The form of the presentation and who will receive it must be agreed. The route cards, tracings and other details must be given to the Supervisor, using the standard *Expedition Notification Form* (the 'green form'), for dispatch to the appropriate Panel Secretary for ventures in Wild Country Areas.

Where a Wild Country Panel Assessor is required notification must be given to the Panel Secretary **at least six weeks in advance**. If the Operating Authority is providing its own Accredited Assessor, notification to the Panel Secretary must be made at least four weeks in advance.

For ventures **outside** the United Kingdom **notice must be given to the Operating Authority at least twelve weeks in advance** using the standard *Notification Form for Ventures Abroad* (the 'blue form'). See Chapter 2.5 - Ventures Abroad.

By the time the notification forms have to be dispatched, whether for the United Kingdom or abroad, fitness training should be well under way and increasing in intensity.

6. Assessors, with their more intimate knowledge of the area where the venture is taking place, may suggest amendments to the proposals, especially concerning the routes and camp sites, and will then inform the Supervisor; this is known as the **Initial Contact**. The Supervisor will, in turn, notify the participants who should then make any necessary modifications to their venture.

7. The **Local Pre-expedition Check** is intended to help the participants by ensuring that their training and equipment is of a sufficiently high standard to embark on the venture. This will avoid the disappointment of travelling all the way to the venture area (wasted effort, time and money) only to find that there are deficiencies in their training or equipment which may prevent the venture from taking place.

   An Accredited Assessor (or someone with considerable experience in the mode of travel, a familiarity with the Award Scheme and the intended area of the venture and who lives locally) meets the group at their home base and carries out a check of their equipment and training. The best time for this to happen is a week or ten days before departure to the venture area. If it takes place earlier, items of equipment which may have to be borrowed, such as tents, may not be available while, if it takes place later, there may not be time to remedy any deficiencies in training or equipment.

8. The next step in the chain of events leading to the qualifying venture is designed to help the participants by meeting with their Assessor.

This is known as the **First Meeting** and takes place during the 48 hours' familiarisation period which is advised for all ventures in wild country, prior to embarking on the journey. For ventures at Bronze or Silver levels in normal rural or open country, the **First Meeting** usually takes place immediately before departure. It is the Assessor's opportunity to put participants at their ease, agree on the ground rules and establish some form of contract on expectations. Communication channels and emergency procedures will be established, the form of the report and who shall receive it will be agreed, and the process of observing and recording will be checked. If a **Local Pre-expedition Check** has not been carried out, the Assessor will wish to confirm that the group is adequately trained and the equipment satisfactory. Sometimes Assessors wish to reassure themselves, even when a **Local Pre-expedition Check** has already been completed. Each participant should be in possession of a fully signed up *Record Book.*

9. For a well trained and equipped group the venture itself should be the culmination of their efforts, the realisation of their ambitions, and should be enjoyed by all. Preparation and physical fitness should ensure success whatever the weather. Regardless of the conditions encountered during the journey, experience indicates that all ventures in the Expeditions Section are memorable and are significant events in the lives of the participants.

The Supervisor, who has the responsibility for the group's safety, usually meets the group en route and in the evening at the camp site. The Assessor, whose role is to ensure that the conditions of the Expeditions Section are fulfilled, will meet the group as frequently as necessary to ensure that the conditions are complied with (this is normally once a day), and write a report in the *Record Book.* Sometimes groups feel that they have not had a sufficient number of visits from the Assessor. They should not feel deprived as Assessors usually have more than one group to attend to at the same time, travel considerable distances and hold down a job all at the same time. Ventures belong to the participants and all visits by adults are intrusions into their venture, even though they are necessary and are a requirement of both the Operating Authority and the Award Scheme. Because of the high mileages involved in Wild Country Areas, it is sometimes necessary to pass groups from one Assessor to another during the progress of the journey.

10. On completion of the journey the Assessor will carry out an oral **Debriefing** with the participants. This is the opportunity for the Assessor to share in the participants' experiences and their reactions to the venture, the ups and downs and the difficulties they have overcome in meeting the challenge, and then to congratulate them on the completion of the journey.

11. When the venture is over, all members of the group have to be involved in producing a presentation. If it is an oral presentation, it may take place when the **Debriefing** has been completed and before the participants depart for home. Special arrangements must be made in advance for this to happen and ample time set aside for the process, with suitable accommodation, event if it is only a minibus. Both the appropriate pages of the *Record Books* may then be signed by the Assessor, completing the requirements for the Expeditions Section.

Where oral presentations are going to be submitted immediately after the Expedition, participants will need to allocate time during each day of the venture to prepare material after observations and recordings have been made.

Where other forms of presentations are going to be submitted, the Assessor will wish to confirm these arrangements once more, and stress the importance of submitting the report. The advantages and disadvantages of the various methods of reporting are discussed in Chapter 3.15 - Observing, Recording and Presentations. The Assessor will sign the relevant page of the *Record Books*. Participants should remember that the greatest number of uncompleted Expeditions arise from the failure to submit a presentation of the venture. They should prepare the presentation as soon as possible after the completion of their venture while the experience is fresh and motivation high.

EXPEDITION GUIDE

Part
6

Ian Shaw / National Trust

SUPPORTING

ROLES

AND

RESPONSIBILITIES

Part 6

Supporting Roles and Responsibilities

Training

Training is the most demanding of the adult roles in the Expeditions Section and Instructors deserve the support of all the other adults involved within the Award Scheme. It demands a commitment throughout the year and most Instructors start training the next group as soon as the previous groups have completed their qualifying ventures. There are as many patterns of training as there are Award Units. Many Units rely on one person who does everything, others have a team of people who share the workload, while some bring in Instructors to deal with specific aspects of the syllabus.

Successful training is based on a systematic approach with clearly defined objectives. Many Instructors involved in Expedition training are teachers, youth workers or those involved in training in commerce and industry. One hesitates to recommend teaching methods to those who have received professional training. There are, however, many involved in the Award who have not received any professional training; the only teaching methods they may be aware of are those which they experienced at school a number of years ago and which may not be appropriate for the Award. The following advice may be helpful.

The Expeditions Section places special demands on Instructors for a number of reasons.

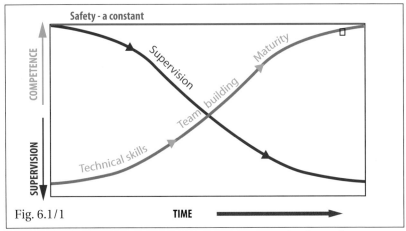

Fig. 6.1/1

Qualifying Ventures are Unaccompanied and Self-reliant

Because participants are not accompanied by Adult Leaders on their qualifying ventures they are totally dependent on their equipment, training and experience for their safety and success. The educational ethos of the Award Scheme is that of self-reliance and independence. To achieve this it is essential that the amount of direct supervision is reduced, gradually and progressively, as the participants' skills, confidence and experience increase. The decrease in directly supervised activity must be balanced by the participants' increased competence and self-reliance so that the high levels of safety, so essential in outdoor activities, are always maintained. Instructors must observe individual participants very closely as direct supervision is withdrawn in sequential stages. This will ensure that participants are able to cope with the reduction in supervision, not only in terms of increased skills, but also in self-discipline, responsibility, behaviour, maturity and judgement.

Participants must be able to Take Charge of their Venture

To enable participants to take charge of their ventures, Instructors must base the training on:

- The philosophy, requirements and conditions set out in the *Award Handbook.*
- Team building and leadership.
- Skills related to living in the outdoors, the mode of travel, safety, the care and conservation of the environment; physical fitness for the venture, observing, recording and reporting.

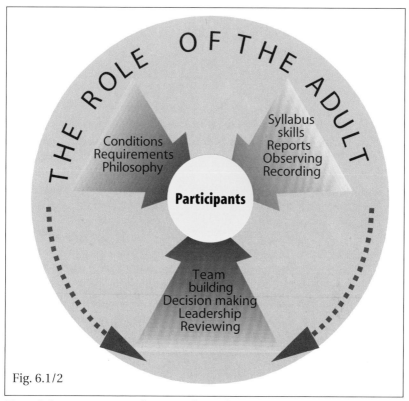

Fig. 6.1/2

Participants must know the Philosophy, Requirements and Conditions of the Award

Success depends on fulfilling the philosophy, requirements and conditions of the Award. Participants need to be as much aware of these as Assessors and Supervisors; this will help to ensure that no action on their part will invalidate the venture and they will be able to fulfil both the specific conditions and the philosophy of the Award. Direct access to relevant literature is essential for all the participants. Failure to provide this information is rather like asking them to play a team game, such as hockey or football, without knowing the rules.

All Ventures in the Expeditions Section are a Team Effort

Ventures depend on the success of each individual member, and mutual support and collective experience are essential to achieve a common purpose. Team building should start from enrolment and be an ongoing process. Groups should be allowed to form naturally from the individual members of the Unit, and friends should, where possible, remain together. The natural bonding between friends is important as it tends to be more enduring in the face of adversity. Groups should be composed of compatible individuals who will support each other to achieve collective success.

Once a group has formed, training should strengthen and weld it together to cope with the inevitable stresses of a venture. Working together in the training programme, cooking for each other, living together in the same tent, being actively involved in real decision-making and accepting the consequences of the decisions, whether right or wrong, forms the base for successful team building. This process is enhanced by internal leadership, by reviewing and monitoring their own progress through the syllabus and training.

Unsuccessful ventures are not usually the result of inadequate technical skills but rather through a lack of group commitment, poor group and personal relationships, or a disregard for the requirements. The strength of the group is particularly important when faced with foul weather or where members are, for instance, badly blistered; only determination and mutual support keeping the venture alive.

The Skills and Techniques of Expeditioning

The Hard Skills

It is customary to refer to all the technical skills involved in navigation, camp craft, first aid, canoeing, walking, biking and other activities and pursuits as the 'hard skills'. Competence in these skills forms the foundation of safety and well-being in the outdoors.

The Soft Skills

The skills of leadership, team building, group and interpersonal relationships - the 'people skills' - are usually referred to as the 'soft skills'; the analogy probably being made with 'computer speak' in the first instance.

Both the hard and soft skills are of equal importance in ensuring a successful outcome in the Expeditions Section of the Award and demand equal attention on the part of the Trainers. It is usually more effective to use the hard skills as a basis, or vehicle, for teaching the soft skills rather than teaching them theoretically or as abstract concepts, but the soft skills should feature specifically in all training sessions and have a special place in the practice journeys and in reviewing.

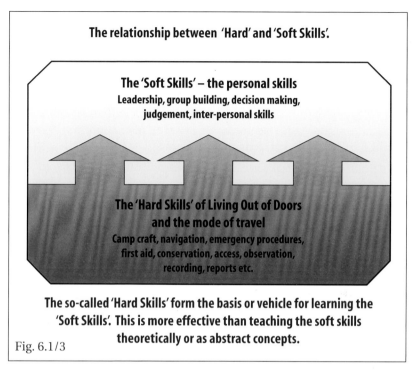

The relationship between 'Hard' and 'Soft Skills'.

The 'Soft Skills' – the personal skills
Leadership, group building, decision making, judgement, inter-personal skills

The 'Hard Skills' of Living Out of Doors and the mode of travel
Camp craft, navigation, emergency procedures, first aid, conservation, access, observation, recording, reports etc.

The so-called 'Hard Skills' form the basis or vehicle for learning the 'Soft Skills'. This is more effective than teaching the soft skills theoretically or as abstract concepts.

Fig. 6.1/3

Conservation and Care of the Environment

The Country Code, and any other relevant codes of practice and behaviour, must always have a specific place in training and their importance should be emphasised. It is vital that care for the environment must permeate all aspects of training so that actions which will lessen the effect of the venture on the countryside or its inhabitants are discussed and implemented, and any actions which will have a high impact on, or be destructive of, the environment may be avoided.

Safety

In outdoor pursuits there are 'non-learning situations' where there is only one way of doing something, **the correct way**, where the consequences of using the wrong technique may have disastrous results. Examples which spring to mind are fuelling and using stoves, throwing a life-line, fastening on to a life-line, capsize drill and resuscitation.

Training in crucial techniques, some simple and basic such as travelling on a compass bearing, should involve such instruction that it becomes a 'drill', so that even when the mind is numbed by cold, or traumatised by shock, response is automatic and the participants are 'programmed' to follow the correct procedure without conscious thought.

A STRUCTURED TRAINING PROGRAMME

A training programme must be prepared which will provide the necessary levels of competence, safeguard the well-being of the participants and ensure that there is complete coverage of the syllabus. A comprehensive scheme of training should be prepared for the Unit, or group, as soon as they embark on a particular level of Award in the Expeditions Section. This is usually done by dividing the available time by the syllabus which has to be covered; this should give the number and length of the training sessions or, if there is insufficient time, the earliest date when the qualifying venture may take place.

Although there is an endless variety of Award Units, each with its own pattern of training, it is possible to discern a common pattern in the training programmes set up by the vast majority of Units. The programme is largely influenced by holidays, the Expedition season and the academic year. Most Units begin training in the Autumn after the completion of the previous groups' qualifying ventures in the Summer. The programme frequently consists of about one and a half or two hour training sessions once a week in the Autumn and Winter with an interval over Christmas and the New Year. Practice journeys start at the end of March and take place into the Summer, along with further training sessions, leading to the qualifying ventures in the Summer or early Autumn.

The majority of all training for the Expeditions Section has to take place out of doors. Instruction involves getting cold and wet as well as drying and cleaning wet clothing and equipment; a fact which all Instructors must come to terms with. There are, however, certain aspects of the syllabus which are more conveniently taught indoors, such as first aid, route planning, some aspects of safety and emergency procedures, the preparatory skills of map-reading, and the codes related to the care of the countryside and the mode of travel. With careful planning it may be possible for training in these aspects of the syllabus to take place indoors in the middle of Winter when the weather is usually at its worst.

About 90% of navigational training will still take place out of doors and the skills of first aid and safety and emergency procedures must be tested by exercises outside.

Those aspects of the syllabus which take the longest time to deliver should be started first. A group intending to change to an unfamiliar mode of travel, say from walking to canoeing, will find it essential to begin training straight away as it may take at least a year to gain the necessary competence. This early training is assisted by the fact that water temperatures in and around the British Isles reach their maximum towards the end of the Summer and the early Autumn. Skill reinforcement may come from other Sections of the Award Scheme; for example, if a water venture was being considered, then Boatwork from the Skills Section or one of the Survival or Safe Swimmers Awards from the Physical Recreation Section would be a most effective way of increasing the depth of experience and widening the margins of safety.

The training programme must take into account the time of the year as well as the availability of suitable Instructors. The practice journeys must be included in the programme and it is advisable to have a little spare time in hand in case training does not go according to plan. Where

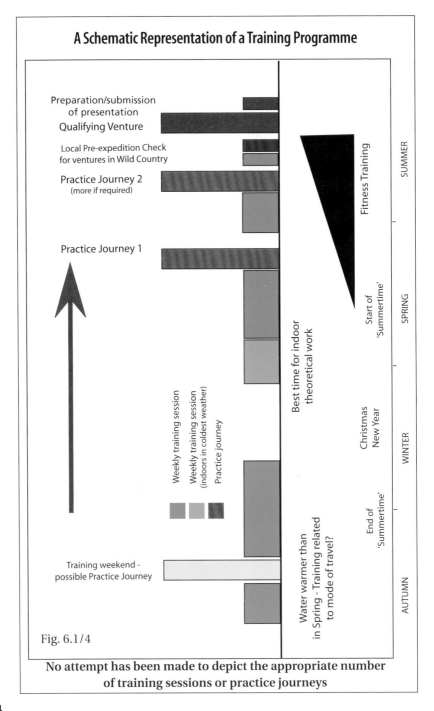

A Schematic Representation of a Training Programme

Fig. 6.1/4

No attempt has been made to depict the appropriate number of training sessions or practice journeys

training is going to take place over the better part of a year it is an understandable human failing to start the instruction in a leisurely manner; this frequently results in a frantic effort to complete the programme as the time for the venture approaches. There must be an urgency, a determination, in all instruction from the very first session which must be maintained until training is completed. One person should be responsible for ensuring that all the contents of the syllabus are covered, and care should be taken to be certain that no elements are omitted from an individual participant's training. A comprehensive checklist should be prepared, based on the syllabus in the *Award Handbook* and the *Expedition Guide*, to assist in this process, committed to paper and made available to the participants. Individual progress throughout the training process can then be checked.

The time taken to become competent in different aspects of the syllabus will vary with the aptitudes and abilities of the participants, but it is still possible for Instructors to make a reliable estimate of the time needed to deliver each aspect of the syllabus. These times, or the number of instruction sessions, should be recorded as they are a most valuable teaching aid. They will serve as a guide for planning future training programmes and enable appropriate portions of time to be allocated to different aspects of the training syllabus. Priority must be given to aspects of the syllabus which the participants find most difficult and which take the longest time to acquire a satisfactory level of competence.

The progressive nature of the Award through Bronze, Silver and Gold levels makes it possible to build on training from lower levels of the Award and enables greater depths of competence to be achieved. Instructors should remember that participants who have completed their Bronze, or particularly their Silver Award, will bring considerable competence and experience with them. This will enable Instructors to include beneficial training weekends or extra practice journeys early in the training programme, which are especially useful where there is a change in the mode of travel from the previous level of the Award. Provision must be made for individuals who are direct entrants at either Silver or Gold level. Systematic revision will prevent omission in the training of direct entrants. In the case of Explorations, it is helpful to find someone early in the training process who is knowledgeable in the chosen field of study to act as the group's Mentor.

PLANNING TRAINING SESSIONS AND PRACTICE JOURNEYS

Just as it is necessary to prepare a training programme from the induction of the group to the qualifying venture, so it is equally essential to prepare a plan for each training session and practice journey. Each training session should be based on the following elements :

- Mixing different aspects of the syllabus.
- Concentrating on the essentials by eliminating unnecessary background material.
- Learning by doing rather than by talking.
- Providing sufficient equipment for everyone.
- Repetition and revision.
- Reviewing.
- Recording.

Sessions benefit from mixing different aspects of the syllabus, say camp craft with navigation. This may involve a subject which is essentially practical with one which is largely theoretical, or an outdoor activity with an indoor activity. Elements of the requirements, conditions and aims of the Expeditions Section should be included as well as the use the practical skills as a means of delivering the soft skills. The combination of too many different elements may lead to fragmentation and a lack of focus.

The essential simplicity of the training syllabus should be retained. There is rarely sufficient time available for the inclusion of unnecessary background material which frequently leads to confusion. For example, a simple six figure map reference will meet all the needs of the participants. There is no need to understand four or eight figure grid references, or how the grid was constructed. The Award is about kindling interest; there will be opportunities later to pursue activities to greater depths of understanding.

Sessions should be built around **doing** by the participants rather than being **talked** at by the Instructor. While a teaching or instructional input is an essential element of nearly all training sessions, the most effective and suitable learning is derived from practical, active participation in the learning process. This is particularly so for the Award with its increasing involvement with older participants.

A common bottleneck in training is a lack of equipment. Where there is a shortage of equipment, it inevitably falls into the hands of the most competent or assertive group members who increase their competence at the expense of the less able. Instructors must ensure that there is enough equipment **for each individual member** of the group. The sharing of maps, compasses, stoves and other equipment is unsatisfactory. **Training must always be directed towards individuals rather than the group.** Any individual deficiency in the basic skills poses a threat to the safety, not only of the individual, but of the rest of the group. Instructors may have to resort to borrowing equipment from other Units, staggering and co-ordinating training and other strategies to overcome a shortage of equipment.

Each training session should contain an element of repetition and revision of material from previous sessions. Repetition should be built into each session and will enable skills to be 'grooved' and become part of the participants' being. In a voluntary scheme, there will inevitably be absences; repetition helps to reduce the possibility of individuals' missing out on essential techniques. Repetition is a vital factor in the learning process and in ensuring the safety of the individual.

Reviewing

Reviewing is an everyday process used throughout education, industry, commerce and administration. The term is used loosely to cover any retrospective survey or consideration of an activity. It is usually performed by those who manage or control. In Outdoor Education reviewing has, in recent years, taken on a more specific meaning and is applied to the process whereby, together with the Instructor, all members of a group examine, consider and reflect on an instructional session or learning experience.

The purpose of reviewing is to clarify and consolidate the learning and enable participants to absorb it and, by discussion, identify any difficulties which they have encountered. A reviewing element built into every session enables participants to examine and analyse what they have learnt and the ground covered; it facilitates revision and the preparation of future training sessions. The process of reviewing with its discussion and analysis significantly reinforces the learning process. It enables participants to reflect on their individual contribution to the group and is an essential part of team building when they discuss the group's strengths and weaknesses. Participants should be involved in future planning to ensure that their personal needs are included.

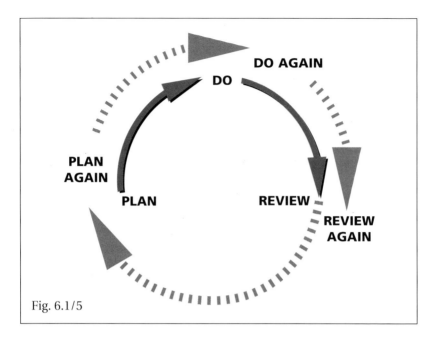

Fig. 6.1/5

Reviewing with the group has particular benefits for Instructors, enabling them to evaluate learning progress and diagnose difficulties experienced by individuals, as well as providing an opportunity to observe group and personal relationships. Involving a group in a regular reviewing process demands skills and a sensitivity on the part of the Instructor.

Reviewing is a relatively simple process, but young people will require assistance in the initial stages of the training programme. Instructors can help by asking open questions which initiate discussion. Reviewing must be based on the 'plan-do-review' diagram above:

- "What did we plan to do?" - the purpose of the practice journey or training session.
- "How well did we achieve these aims and objectives?" - ample illustrations of the questions leading to discussion and evaluation will be found in Chapter 6.2 - Effective Practice Journeys.
- "Some of us ran out of methylated spirits. How much should we take next time?"
- "What was our speed of travel this afternoon?"
- "How could we increase this in the future?"

Later questions are best directed to the soft skills where more sophisticated value judgements have to be made:

- "Why did the group become so spread out in the morning?"
- "Was this a problem of not working together as a team, or a matter of leadership?"
- "How can this problem be overcome in the future?"

The reviewing process is successful when participants can initiate their own reviewing without adult intervention.

The Award Scheme's publication, *Playback - A Guide to Reviewing Activities*, provides a wealth of information and ideas help to build reviewing into training sessions. Reviewing must never become an end in itself; its purpose is always to reinforce the learning process and it must only occupy a small proportion of the time available.

Recording is an integral element in the learning process. Instructors should keep a record of when the various aspects of the training syllabus are completed so that they, or the Supervisors, may sign the participants' *Record Books* knowing that there have been no serious training omissions which may affect the safety of those involved. Recording of past work forms the basis for planning future training. Individual participants should log their own training progress as part of the reviewing process.

Practice Journeys

Practice journeys, a mandatory requirement, are detailed in the *Award Handbook* and play a significant role in the training programme. They provide peak experiences enabling isolated technical skills to be combined and provide the greatest opportunity to exercise the soft skills and undertake the reviewing process. During practice journeys Instructors should always include an unexpected, or surprise, short practical exercise on some aspect of first aid or safety and emergency procedures which will reinforce the theoretical instruction. The importance of practice journeys is recognised by the inclusion of Chapter 5.2 - Effective Practice Journeys.

The Training Session

Using the syllabus in the *Award Handbook*, the *Expedition Guide* and the advice outlined above, it is possible to devise written notes for each and every training session. Each session will normally incorporate the following, though not necessarily in a particular order:

- Revision of previous work.
- A teaching input and/or demonstration by the Instructor.
- Individual and group practical work.
- Provision for discussion, review and recording.
- Preparation for future sessions and assignments.

The plan should list the equipment and materials needed, access to any facilities required and transport. Participants should always be informed of clothing, dress or equipment requirements well in advance so that they may be properly attired for each session, which will nearly always involve activity out of doors.

To be successful, training sessions must be:-
- **Enjoyable.**
- **Stimulating.**
- **Challenging.**
- **Practical.**
- **Relevant to the participants' needs.**
- **Based on the participants' existing knowledge.**

Training sessions may be evaluated in the reviewing process against these criteria.

EXPERIENCE AND JUDGEMENT
Experience
The gradual and progressive acquisition of experience is second only to training in importance. Unless training is consolidated by experience there is a great danger that much of it will be wasted. A carefully planned training programme with ample repetition and revision, coupled with the appropriate practice journeys, forms the best basis for the acquisition of experience, but additional involvement by the participants is required. Experience comes largely through time and variety. Diverse situations are needed if the greatest benefit is to be derived from the commitment of time and effort, and participants should be encouraged to involve themselves in activity outside the formal training sessions. The benefit of this extra-curricular activity is all the more valuable if it can involve a structured programme of physical training in preparation for the qualifying venture.

Judgement

Of all the qualities entailed in the safety and well-being of participants in Award ventures, that of sound judgement is the most important. Sound judgement, along with responsibility and maturity, arises from effective training coupled with progressive and varied experience over a period of time. It cannot develop unless there are opportunities to exercise judgement. The Expeditions Section is all about providing opportunities to exercise individual and collective judgement.

Joe Cornish

Supervising Qualifying Ventures

A ll ventures, including practice journeys, must be supervised by a suitably experienced adult, who must accept responsibility for the safety and well-being of the group on behalf of the Operating Authority. The Supervisor, who is the agent of the Operating Authority, must be satisfied that the participants are fully trained and equipped to undertake the planned venture.

Supervision is necessary for both qualifying ventures which are assessed, and for the practice journeys which are so much part of the training process that they are dealt with elsewhere in Chapter 5.2 - Effective Practice Journeys.

Groups undertaking their qualifying venture **must not be accompanied** by an adult except in very special circumstances when the Operating Authority may permit closer supervision. Exceptions are made for some water ventures. In certain circumstances, for safety reasons, it may be important that a group is able to make contact with an adult at night.

Supervisors who have not established a base in the immediate vicinity of the venture must be sufficiently close to render help within a reasonable time if an emergency should arise.

Supervisors:
- Should be familiar with the aims and objectives of the Expeditions Section, and with the conditions and requirements which the participants have to fulfil.

- Should be sufficiently experienced, competent in the mode of travel and the skills of navigation to be able to provide safe and effective supervision.
- Should ensure that the parents have been informed of the unaccompanied and self-reliant nature of Award ventures and the mode of supervision.
- Should be present in the area of the venture for those in normal, rural or open country.
- **Must be based in the area for ventures which take place in Wild Country Areas.** This requirement extends to Gold Explorations on the sea coast and remote areas of marshland.
- **Must be based in the area for all water ventures.** See Part 4 - Advice and Skills Associated with the Mode of Travel.
- Should not be involved in activities or work or have responsibilities which would prevent them from rendering urgent and effective assistance to the participants.

The majority of Operating Authorities have their own Regulations and Codes of Practice which must be observed.

The demands made on Supervisors, in terms of technical knowledge and experience, tend to be greater in ventures which take place in wild country and on water, but all ventures which involve the care of young people out of doors require anticipation, alertness, vigilance and care from all who have responsibility for safety.

With greater numbers of entrants in the range of 18 to 24 years of age, many participants are adults in their own right. The supervision of these adults would not need to be as close as that of a group of 14 year olds embarking on their Bronze venture. There is, however, a responsibility to ensure that all participants are correctly advised and the responsibilities of care extend even to the 24 year olds.

The nature of supervision varies greatly between Operating Authorities. Many Supervisors are Unit Leaders or are involved in Expedition training, while others are involved in assessment. Some Operating Authorities have Expedition Teams where the members are allocated to supervisory or assessment roles on different occasions. To avoid confusion the title 'Expedition Panel' is reserved exclusively for Wild Country Expedition Panels operating in the Award Scheme's designated

Wild Country Areas, and for a few non-Wild Country Panels, based in areas such as Exmoor, the Forest of Dean and the New Forest, supported directly by The Duke of Edinburgh's Award.

Supervision can be a lonely and daunting task, especially when a group is overdue and the weather is closing in. Such situations call for a fair amount of 'bottle' on the part of the Supervisors if they are not to be panicked into taking premature action and creating unnecessary alarm. Groups are frequently overdue at their destinations but they eventually turn up tired and weary. Occasionally they do not reach their destination but, by carrying their camping gear, they are able to camp short of their objective in perfect safety. Judgement based on previous experience is always of great help in such situations. Most Supervisors work closely with the Assessor who can provide advice and support.

The Supervisor, the group and the Assessor are all part of a team formed to bring about a successful conclusion to a venture which fulfils the requirements and conditions of the Expeditions Section. Supervisors need to be just as familiar with the requirements as the group and the Assessor, and must be careful not to take any action contrary to the conditions which would invalidate the journey. If a group is very late reaching a camp site it would be wrong to give them a lift in a mini-bus as this is contrary to the conditions - better for the group to camp where they are and make up the distance over the rest of the venture, remembering, of course, to inform the Assessor.

The Supervisor is the focal point of communication before, during and after the venture and must keep everybody informed of any changes of plan, especially if alternative routes are used in foul weather, or if the venture is aborted. There have been far too many instances of Assessors trying to locate a group in the middle of the night, when it is safely tucked up in bed a hundred miles away. It is most important to have access to a telephone or, failing this, the ability to receive a message promptly. There will inevitably be an increase in the use of mobile phones over the next few years. This may well facilitate the supervision of Bronze and Silver journeys in normal rural or open country, but mobile phone users should be aware that the coverage in mountainous areas is at best patchy and often non-existent. (See Chapter 6.3 - Mobile Phones and Radios).

Ventures belong to the participants and any intervention, other than the minimum necessary to ensure the safety of the group and that the conditions are fulfilled, is an intrusion into their venture. When groups have to be shadowed by the Supervisor it is nearly always an indication that the participants are inadequately trained or that they are engaged in a venture which is beyond their competence, in an environment which may be too difficult for them. The completion of a prescribed distance or duration is only a means to an end. The prime aim of the venture is to make the participants self-reliant and work together as a team, making real judgements and decisions.

The Supervisor's role is certainly as important as that of the Assessor and the quality of the experience can be enhanced or marred for the participants by the Supervisor just as easily as by the quality of the assessment.

The duties of the Supervisor, detailed below, are for ventures at Gold level in wild country, but will serve as a pattern for good and safe supervision of Bronze and Silver ventures and for those using all modes of travel. The references below in bold refer to significant stages in the assessment process for Wild Country Areas and are defined in Chapter 6.4 - Assessment.

GUIDANCE NOTES FOR SUPERVISORS
Before the Venture

- Verify that the training has been completed and the participants are properly equipped.
- Ensure that the *Record Books* have been signed.
- Check that the notification forms, route cards and tracings are complete, including the nature of the presentation and to whom it will be submitted, and sent them to the Wild Country Panel Secretary.
- Find a suitable base from which to supervise the venture with access to a telephone or where messages can be received from the participants or Assessor.
- Respond to the Assessor after the **Initial Contact** has been made.
- Complete arrangements for the **First Meeting** immediately prior to the venture.
- Collect all relevant safety information which may be required. This will include the names, addresses and emergency contact numbers for the participants' parents/guardians, the Assessor, the responsible person in the Operating Authority and at Award Headquarters.
- Participate in the **First Meeting** along with the Assessor and group during the day or evening immediately prior to the venture.

During the Venture

• Visit the group once a day, or as the needs of safety demand. Visits should be co-ordinated with the Assessor to ensure that the group is visited at the camp site each evening.

• Be responsible for communications and keeping everyone informed of any changes in plans.

After the Venture

• Be present at the **Oral Debriefing** carried out by the Assessor if invited by the group.

• Receive the presentations of the venture if the participants decide to submit them to the Supervisor.

• Check on the progress and production of the reports if they are to be received by someone other than the Supervisor.

ACCIDENT AND EMERGENCY PROCEDURES FOR SUPERVISORS

Many Operating Authorities have their own accident and emergency procedures. In the event of an accident or emergency Supervisors should, in the interests of all concerned, note times and keep a written record of events.

Supervisors receiving a call for help from a group should:

• Ask for all the information required in the emergency message and tell the callers what to do. This usually means staying by the telephone.

• Summon any emergency service if required.

• Notify the Assessor.

• Proceed to the location of the incident to support the group.

• In the event of a serious incident, notify the responsible person within the Operating Authority, and/or others as deemed necessary, including The Duke of Edinburgh's Award emergency number at Award Headquarters.

Supervisors should always have on their person the following information together with an ample supply of coins for the telephone, a phonecard and/or a telephone charge card:

• The names, addresses and emergency contact telephone numbers for each participant.

• The telephone number and address of the Assessor.

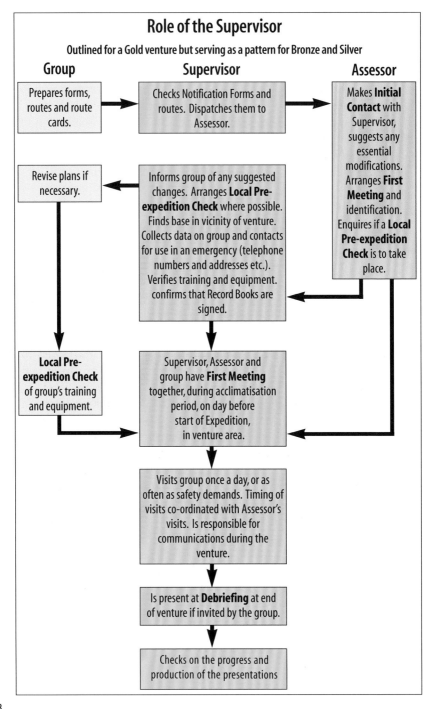

Role of the Supervisor

Outlined for a Gold venture but serving as a pattern for Bronze and Silver

Group	Supervisor	Assessor
Prepares forms, routes and route cards.	Checks Notification Forms and routes. Dispatches them to Assessor.	Makes **Initial Contact** with Supervisor, suggests any essential modifications. Arranges **First Meeting** and identification. Enquires if a **Local Pre-expedition Check** is to take place.
Revise plans if necessary.	Informs group of any suggested changes. Arranges **Local Pre-expedition Check** where possible. Finds base in vicinity of venture. Collects data on group and contacts for use in an emergency (telephone numbers and addresses etc.). Verifies training and equipment. confirms that Record Books are signed.	
Local Pre-expedition Check of group's training and equipment.	Supervisor, Assessor and group have **First Meeting** together, during acclimatisation period, on day before start of Expedition, in venture area.	
	Visits group once a day, or as often as safety demands. Timing of visits co-ordinated with Assessor's visits. Is responsible for communications during the venture.	
	Is present at **Debriefing** at end of venture if invited by the group.	
	Checks on the progress and production of the presentations	

- The telephone number and address of a responsible person within the Operating Authority.
- The telephone number and address of the Unit Leader, or Head Teacher/Principal Youth Officer/Works Manager as appropriate.
- The telephone number of the local doctor and the location of the nearest hospital with a casualty department - the majority of accidents do not necessitate the use of a rescue team or the emergency services. Mountain rescue teams are alerted by a 999 call via the police.
- The Duke of Edinburgh's Award's emergency telephone number for use in the event of a serious incident (01753 727400).
- Any parental consent forms required by the Operating Authority for medical treatment.

When rescues make headline news all parents are concerned in case their sons and daughters are involved, and it is frequently the task of the Supervisor to put minds at rest. Only one young person in one group may be involved, but there could easily be twenty or more Award groups involving a hundred or more participants in the same area.

In the event of a serious injury or incident, it is the Supervisor's task to ensure that the parents or next of kin of the victim are informed immediately, and then to notify the parents of the other members of the group that they are safe and well. The police may do this, or assist in this process, or it may be carried out by the Operating Authority's **responsible person** following its own procedures. The police in this country will not normally release the name of a casualty until the next of kin have been informed. **It is essential that the parents are notified before the media are able to make contact with them**. This task may be more easily performed by the **responsible person** or one of the other people listed above, but the task is of such importance that there must be no confusion over who does what. All the key people **must** be informed.

The Supervisor should then protect all the participants from media pressure and, if necessary, request the assistance of the police to do so.

Supervisors must be aware of the concerns and anxieties of the parents of the injured participant and render every assistance and support to them.

Because of the extent of shock and anxiety which is always involved in a serious incident, any response to the media should always take the form of a prepared statement by the **responsible person** rather than the Supervisor involved in the incident. If Supervisors cannot avoid making a statement to the media they should always prepare a written statement beforehand, read the statement, not enlarge on it and not answer any questions afterwards as remarks can be quoted out of context.

There have been very few serious accidents involving participants in the Expeditions Section and it is hoped that this will continue to be the case. Young people are at much greater risk while playing team games or travelling to or from work or school. It is a great tribute to the quality of supervision over the years that millions of young people have participated in Expeditions in safety and the Award Scheme has such a good safety record. Supervisors play a vital role in maintaining this level of safety.

Mobile Phones and Radios

THE USE OF MOBILE PHONES AND RADIOS BY PARTICIPANTS DURING THEIR EXPEDITIONS

The Award Scheme does not recommend that participants carry and use mobile phones and radios during their venture, especially in Wild Country Areas, for the reasons given below. Where mobile phones are carried by the participants the following must be observed.

1. The venture must be planned, supervised and assessed in the normal manner as if mobile phones or radios were not being carried, so that the use of this equipment provides additional emergency protection rather than leading to a diminution of normal practice. On no account must routes be determined by the location of telephone boxes.

2. Radio or telephonic communication must not replace the personal daily contact which the Supervisor makes with the group. Supervisors must observe the conditions relating to their proximity to the venture. See Chapter 6.2 - Supervising Qualifying Ventures.

3. There must be no reduction in the quality of training and the equipment of the group.

4. Mobile phones or radios carried by the participants during their venture should be placed in a package which will be sealed and signed by the Assessor, before being protected from damage by water. **The apparatus must not be used by the participants except in the case of an accident or emergency**. If the apparatus is used the

participants will inform the Assessor at the earliest opportunity. The Assessor will decide if the use of the phone or radio was warranted by the nature of the emergency. Should the Assessor decide that the use of the equipment was sensible and wise considering the nature of the incident, the Assessor will reseal the equipment and allow the venture to proceed if that it possible.

5. Care should be taken to ensure that the use of mobile phones or radios does not give rise to a false sense of security in any venture, especially in Wild Country Areas where coverage is at the best patchy and, over large areas, non-existent.

The most important argument against the use of mobile phones and radios in the Expeditions Section is a philosophical one - they destroy the group's sense of isolation, self-reliance and dependence on their own resources. One of the principle aims of the Expeditions Section is to develop self-reliance, effective decision making and perseverance. This cannot occur when the participants can telephone or radio for advice and ask someone else to make their decisions for them. It may be compared to having the participants on a piece of string or at the end of a long lead.

Jennie Hills (Halina/Fuji Bursary)

There is another strong argument against the use of radios and mobile phones in wild country - for most of the time they will not work. The wave bands on which CB radios operate will, at best, give 'line of sight' communication; mobile phones frequently will not even provide this coverage. While the major national providers claim to cover 98% of the population, this must not be confused with a 98% coverage of the land area of the British Isles, even though the network coverage is increasing steadily. The mobile phone network has to be supported by a radio network of repeater stations which have the limitations of all radios operating on FM frequencies. **Mobile phones cannot be used in the greater proportion of the terrain of the designated Wild Country Areas where land based Gold ventures must take place.**

THE USE OF MOBILE PHONES BY SUPERVISORS AND ASSESSORS

A mobile phone is an excellent facility for all those engaged in supervising and assessing Bronze and Silver ventures in normal rural and open country. They may have limited use in the supervision and assessment of Gold ventures in Wild Country Areas and coastal waters.

A mobile phone removes one of the great disadvantages of the Supervisor using a base camp by avoiding the need to negotiate access to a telephone with the site owner. Establishing a base camp in a location where a mobile phone can be used enables Supervisors to place themselves at the centre of communications, which is one of their most important functions during a venture. Supervisors can have two-way contact with Assessors, parents, Operating Authority, school or workplace. Groups can make contact with the Supervisor from public telephone boxes and, in an emergency, from a farm. The ability of a Supervisor or Assessor to request medical assistance or call on the emergency services is helpful in ensuring the welfare of the participants.

About 85% of all land ventures take place in normal rural or open country where communication is much more dependable than in Wild Country Areas. The range of mobile phones also covers a considerable amount of inland water as well as large areas of coastal waters, extending several miles out to sea over much of the coastline which includes the vast majority of waters used for ventures afloat.

Using a mobile phone in one of the designated Wild Country Areas is unpredictable. Contact may sometimes be established from upland

Jim Hallett, National Trust

areas and in some of the major valleys. It may be possible to establish a base camp which is central to the area of the venture and have communication with the outside world. Groups may be able to establish contact from a public telephone box or a farm. The only way to find out is by experiment, trial and error.

Adult Helpers in the Expeditions Section wishing to purchase a mobile phone for business or domestic purposes should give careful consideration to their choice so that they can maximise its use in the supervision and assessment of ventures at little or no extra cost. A phone which can be charged in or out of a vehicle, or has an alkaline cell adaptor, is a great advantage. Many digital phones have nickel/metal hydride batteries which do not suffer from 'memory' problems to the same extent as nickel cadmium cells, and some phones are able to take larger batteries which increase the 'stand-by time'. Modern mobile phones are highly sophisticated with message retrieval systems and other facilities. Take advice and shop around!

General Considerations

The use of mobile phones and radios within the Expeditions Section is so controversial that it can always be relied on to generate a lot of feeling. The arguments used for and against them are similar to those involved in Citizen Band Radios (CB radios) some twenty years ago. At that time the Award Scheme carried out extensive trials of CB radios in mountainous country to evaluate their usefulness, even though the results of the trials were predictable from the experience of mountain rescue teams. Communication was poor and, at best, would give direct line contact over short distances. In recent years CB radios have been rarely used by participants in Wild Country Panel Areas.

Where ventures last several days the batteries may be flattened. Nicad batteries discharge at a constant and predetermined rate even when not in use and, while this is not apparent when they are kept on their chargers, it soon becomes obvious when they are not able to be charged regularly. This equipment failure increases anxiety which would not exist if these instruments were not being used.

There have been instances of Supervisors relying on Bronze and Silver participants ringing the Supervisor's home instead of the Supervisor visiting the group on a daily basis. When the Supervisor has not been at home, messages have even been left on an answering machine. **This is**

not acceptable. A telephone call cannot provide the same depth of understanding of a group's physical condition, spirit and morale as the discussion, review and observation which occurs when a Supervisor visits the group.

Known abuses of mobile phones include members of groups, using normal rural and open country, ringing home and friends each day and, on some days, on more than one occasion. While the participants' wish to reassure anxious parents is understandable, it is an undesirable use of these instruments and negates the aims and objectives of the venture. Other groups have rung Supervisors for advice and directions repeatedly throughout the day. This would indicate the same inadequacy and lack of training such as where a group has to be shadowed throughout the day by the Supervisor.

The unreliability of coverage and its unpredictable nature in wild country considerably increases the anxiety of Supervisors and Assessors when communication failures occur. This anxiety has frequently resulted in mountain rescue teams being called out when groups have failed to make contact. In many of the reported instances to date, when contact has been made with the group, they have been in excellent condition, in fine spirits and not in need of assistance. The concern and the subsequent calling out of the rescue team has arisen, solely out of the failure, or shortcomings, of the technology involved. **This is an inexcusable waste of the mountain rescue team's time**. It would have been avoided if the instruments had not been used.

Even more disturbing incidents have occurred when groups have called out mountain rescue teams, via the 999 procedure, just because they were lost. It is a basic principle that, when a group gets into difficulties then it is up to the group, and the group alone, to extricate itself from the situation which it has contrived for itself. If there is an accident, illness or injury to the party, then it is an entirely different situation and rescue teams expect to be involved, but for a group to get lost, throw in the towel, and then just to sit down and wait to be rescued is unforgivable. It also ignores the problem of expecting the rescue team to find them when they have no idea themselves where they are.

Parties of hillwalkers, though not necessarily involved with the Award Scheme, are making contact by mobile phone directly with the emergency services on 999. All mountain rescue call-outs go through the police, who ask all callers to stay by the telephone so that the leader of the rescue team can obtain extra information concerning the location and nature of the incident. This has frequently proved impossible with callers using mobile phones who have changed position, sometimes by only a matter of a few metres, making recall impossible.

In spite of the problems mentioned above, there are sound reasons for using mobile phones within the Expeditions Section of the Award, providing they are used in a controlled and responsible manner. The number of mobile phones will increase dramatically with millions more users; Adult Helpers within the Award Scheme will have their share. The amount and quality of coverage over the British Isles will increase steadily as more repeater stations are constructed and, with the third generation of mobile phones, the use of satellites will eventually provide reliable universal coverage.

There may be occasions where the use of mobile phones by the participants could contribute to the safety of the group. The progress of technology cannot be halted or ignored. It is essential, however, that action is always taken to maintain the self-reliant, unaccompanied nature of the Expeditions Section which is so fundamental to its philosophy of the Award Scheme.

Assessment

$\mathbf{A}$ll qualifying ventures within the Expeditions Section must be assessed to ensure that each participant has fulfilled all the relevant requirements and conditions. The principal role of the Assessor is to ensure that all the requirements and conditions, as stated in the *Award Handbook* have been fulfilled. Through this process of validation, the quality and integrity of the Expeditions Section is maintained. The *Award Handbook* defines who can fulfil this role:

- The Assessor must be an adult.
- At Bronze level the Assessor must not have been involved in any of the training or instruction of the group.
- At Silver level the Assessor should be totally independent of the Award Unit.
- At Gold level the Assessor **must** be totally independent of the Award Unit and not associated with it in any way.

For Gold assessments taking place in a designated Wild Country Area, an Assessor from the appropriate Wild Country Panel should be used. Where Operating Authorities do not use an Assessor from a Wild Country Panel, they should approve the appointment of an Accredited Assessor or a person of equivalent competence, preferably holding an appropriate national qualification. Every effort should be made by Operating Authorities to accredit those who have responsibilities for Gold assessments through the Assessor Accreditation Scheme at the earliest opportunity.

The Award Scheme, in conjunction with Wild Country Panels and Operating Authorities, provides a National Accreditation Scheme for Wild Country Assessors. (See Chapter 6.5 - The Assessor Accreditation Scheme). The Assessor's Log Book provides a detailed account of the role, duties and functions of the Assessor.

In addition to ensuring that the conditions of the Expeditions Section of the Award are fulfilled, Assessors have two other functions:
- To advise on the safety of the venture, though the ultimate responsibility rests with the Supervisor who is the agent of the Operating Authority.
- To safeguard the interests of the Award Scheme in the area where the venture takes place.

To carry out their role effectively Assessors must be:
- Familiar with the conditions of the Expeditions Section.
- Competent in the skills common to all ventures as set out in the *Award Handbook.*
- Competent in the skills associated with the mode of travel used by the participants.
- Familiar with the area in which the venture takes place.

Assessors should be in possession of up-to-date *Landranger* maps and any relevant *Outdoor Leisure/Explorer* maps of the area(s) in which they operate.

THE REQUIREMENTS AND CONDITIONS FOR ASSESSMENT

There are some 15 requirements and conditions, 16 if the completion of the journey is included, which have to be fulfilled in the Expeditions Section in order to complete the venture successfully. Expressed briefly, they are:

1. The venture must have a purpose.
2. The group must be responsible for the planning and organisation of the venture.
3. The group must consist of between 4 and 7 young people.
4. The participants in the same group must not be under assessment for different levels of the Award and all members of the group must be within the qualifying ages.
5. The group must not contain anyone who has already completed the Expeditions Section at the level of Award in question, or at a higher level.
6. The environment must be appropriate for the purpose of the venture and unfamiliar to the participants.
7. The venture must take place for the required number of days.
8. Accommodation must be by camping.*
9. The venture must cover the minimum distance, or the required number of travelling hours.
10. The required number of practice journeys must have been completed.
11. The venture must take place during the Expedition season between the end of March and the end of October.*
12. All group members must be properly equipped.
13. All group members must be properly trained.
14. At least one substantial meal should be prepared each day.
15. The journey must be completed
16. A presentation of the venture must be submitted.

* *There are certain exceptions in very special cases - see the Award Handbook.*

The Assessor is a member of a partnership consisting of the participants, the Supervisor and the Assessor, formed to bring about a successful outcome to the group's venture. Participants should regard the Assessor as a person who is as much concerned as they are with the successful completion of the venture - provided that the requirements and conditions are fulfilled.

The task of the Assessor varies with the environment in which the venture takes place, the form of the venture and the mode of travel. **The practices and responsibilities detailed below are for an Assessor operating in a Wild Country Area at Gold level, but it should serve as a statement of good practice for all Assessors at both Bronze and Silver levels and in all environments.**

The first eleven conditions can be checked from the *Expedition Notification Forms* and the equipment and training at a **Local Pre-expedition Check** to ensure the viability of the venture before the participants set out.

The role of the Assessor is much greater than that of checking a list of criteria. Each year many groups owe their success to the wise advice, inspiration and encouragement of their Assessors. Their detailed knowledge of the area and local conditions enables them to make a vital contribution to the success of the venture, as well as advising the Supervisor and the group on matters of safety. This judgement and experience has been acquired through involvement with previous Expeditions and a long association with the area. Guidance, while encouraging and enriching the experience, must never be intrusive or time-consuming and should always be given at the appropriate moment. Assessors must always consult with the Supervisor and be mindful that the venture belongs to the participants.

THE ASSESSMENT PROCESS

The foundations of a successful Expedition are established many weeks before the participants undertake their journey, the Unit or Group Leader, the Supervisor and the Operating Authority having ensured that the venture is correctly set up in accordance with the conditions in the *Award Handbook*.

Assessment should be regarded as 'a continuous process with a number of significant stages' which begins with the Wild Country Panel Secretary receiving the green *Expedition Notification Form*, route tracings etc. and allocating a notification reference number, and ends with the Assessor signing the *Record Books*.

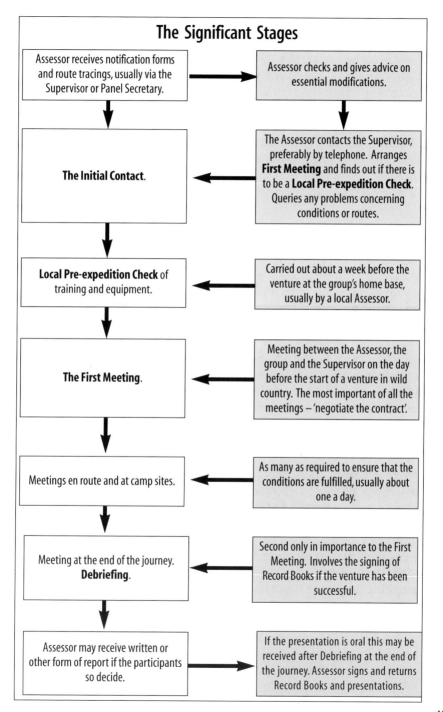

The Significant Stages

Assessor receives notification forms and route tracings, usually via the Supervisor or Panel Secretary.

Assessor checks and gives advice on essential modifications.

The Initial Contact.

The Assessor contacts the Supervisor, preferably by telephone. Arranges **First Meeting** and finds out if there is to be a **Local Pre-expedition Check**. Queries any problems concerning conditions or routes.

Local Pre-expedition Check of training and equipment.

Carried out about a week before the venture at the group's home base, usually by a local Assessor.

The First Meeting.

Meeting between the Assessor, the group and the Supervisor on the day before the start of a venture in wild country. The most important of all the meetings – 'negotiate the contract'.

Meetings en route and at camp sites.

As many as required to ensure that the conditions are fulfilled, usually about one a day.

Meeting at the end of the journey. **Debriefing**.

Second only in importance to the First Meeting. Involves the signing of Record Books if the venture has been successful.

Assessor may receive written or other form of report if the participants so decide.

If the presentation is oral this may be received after Debriefing at the end of the journey. Assessor signs and returns Record Books and presentations.

THE SIGNIFICANT STAGES

Receiving the Expedition Notification Forms and Route Tracings

Wild Country Panel Secretaries will issue a notification reference number and allocate an Assessor to the venture. Assessors should check that the first eleven conditions are fulfilled from the notification form and tracings. They should not suggest changes to a route unless it is absolutely essential for reasons of safety, or because it does not comply with the conditions. Route cards and tracings represent a major undertaking by those concerned involving many hours of hard work. Remember that groups are advised to **go through wild country rather than over**.

Routes which involve around 500 metres of ascent each day should be viewed with suspicion unless it is known that the group is very fit. Late arrivals at camp sites and the exhausted state of many participants is a continual reminder that the vast majority of groups provide themselves with challenges near to the limit of their stamina. Expeditions on foot must cover a specified distance so the physical demands of a journey are best regulated by limiting the amount of climbing.

The Award Scheme does not approve of the use of long distance footpaths but, with the increasing number of these paths in some regions, the use of some sections may be necessary to link one area of country with another.

The Initial Contact

After checking the notification form, Assessors should contact the group's Supervisor, preferably by telephone, carry out the necessary introductions, approve the submission or suggest amendments. They should arrange to receive the group's route cards, equipment lists, menu plans and further details of the venture's purpose if they have not received them already. Assessors should check if a **Local Pre-expedition Check** is to be carried out and, if the venture is an Exploration, whether a Mentor is involved. A convenient time for the **First Meeting** should be arranged during the familiarisation period in the Wild Country Area on the day or evening before the venture starts, when the Assessor, group and Supervisor can all meet together. Means of identification should be exchanged, such as vehicle descriptions, registration numbers and location.

The Local Pre-expedition Check

Local Pre-expedition Checks are intended to prevent participants travelling hundreds of miles to a Wild Country Area only to find that their equipment or training is inadequate. This check should enable conditions 12 and 13 to be satisfied. It should take place at the group's home base normally 7 to 10 days before departure; if earlier, the group may not be able to assemble all their equipment and if it is later, there may not be sufficient time to remedy any deficiencies in training or equipment. The check should be carried out by an Accredited Assessor or an experienced person delegated by the Wild Country Panel or Operating Authority. The check must be carried out by oral questioning, demonstrations by the participants and visual inspection. Written tests should not be used as they are inappropriate. A friendly, informal relationship must be established and the group should not be lectured. The check should last under an hour, with the group doing the talking.

Personal equipment and clothing, emergency equipment and the group's camping equipment should be checked using the advice given in this *Expedition Guide*. Where the absence, or adequacy, of any particular equipment is questioned, it should be considered in relation to the overall level of provision. If a pair of trousers is regarded as inadequate, the garment, when related to track suit bottoms and a good pair of waterproof over-trousers also being carried, may then be perfectly adequate. Many young people may have borrowed their equipment. The ability to keep equipment dry, especially sleeping bags and spare clothing, is important. Pack weights should be checked - bathroom scales are enormously helpful.

Concentration on the practical map-reading skills rather than on the preparatory skills is important. (See Chapters 3.6 and 3.7). The ability to travel on a compass bearing, except at Bronze level where it is usually unnecessary, should always be checked, as should the awareness of the dangers associated with the various forms of fuel and stoves.

There may be wide variations in ability but, providing individuals have the necessary basic competence and are able to carry out the venture without being a danger to themselves or a hazard to the rest of the group or the environment, attention should be directed towards the overall competence of the group.

The First Meeting

This is by far the most important meeting between the Assessor, the group and the Supervisor and may play a vital role in the outcome of the venture. It should take place in the Wild Country Area, during the familiarision period, on the day or evening before the venture commences.

Assessors should establish friendly relations and remove any fears and apprehensions which the group may have concerning Assessors. They should agree on the conditions and establish a 'contract' on what is involved and the expectations of the Award.

Assessors should again confirm the nature of the presentation and to whom it will be submitted. Where oral presentations are to be made to the Assessor at the end of the venture, arrangements should be verified.

Details of the route, the exact location of camp sites, alternative routes for foul weather and escape routes should all be checked. The timings on route cards are frequently too optimistic and attention may need to be drawn to this. A revision of the route card timings can be made at the camp site at the end of the first day.

If a **Local Pre-expedition Check** has not taken place, the Assessor should carry out a check in accordance with the advice given, though the opportunity to remedy any shortcomings at this late stage is extremely limited. It is always worthwhile to check the waterproofing of equipment and pack weights. Bathroom scales carried in the boot of the car are useful for this purpose.

The Supervisor and the Assessor should establish where they will be based and the means of daily communication in case anything should go wrong. Since all three parties are on the move, this is usually achieved by telephoning an agreed number which is likely to be manned. Decide on the action to be taken in the case of an emergency or if the venture has to end through illness or impossible weather conditions.

Finally, the Assessor and Supervisor should agree on how many visits are to be carried out and when. The Supervisor, who has the responsibility for the safety of the group, will wish to make contact with the group at least once each day. By co-ordinating the timing of their visits, at different times of the day, contact can be increased and spread more evenly throughout the day without increasing the total number of visits.

Groups have expectations of Assessors, and Assessors should endeavour to meet these expectations, even though the groups' expectations may be set too high. Assessors should explain to the group that the venture belongs to them and that one visit a day will probably be the norm, except at the beginning of the venture, if all is going well. Where a team of Assessors is working together with a number of groups and more than one Assessor will be involved in the assessment, this should be clearly explained. Frequently it is necessary to pass groups from one Assessor to another if the venture covers a considerable distance or spans more than one Wild Country Panel Area. Many groups fail to realise that Assessors have full time jobs, may be implicated with the assessment of other groups and involved in many hundreds of miles of travel. This will be appreciated by the participants if it is made clear to them before the start of the venture.

Meetings During the Venture

Assessors should make contact with the group as often as is necessary to ensure that the conditions are fulfilled. This should normally entail one visit a day. If the group has choosen to present an oral report at the end of the venture, the Assessor should take the opportunity, when visiting the group, to monitor the preparation of this report. All meetings are an intrusion into the group's venture and tend to destroy the sense of remoteness and self-sufficiency which is central to the philosophy of the Expeditions Section. Visits should be varied between meetings en route and at camp sites in the evening or morning

Groups will frequently be late at checkpoints, and usually at their camp sites, but Assessors must come to terms with this. The majority of participants have insufficient mountain walking and backpacking experience at this stage of their development to make reliable forecasts of journey times. Assessors know that groups progress across the country at around 2.5 kph, or one and a half miles an hour, taking at least 8 hours for a 20 kilometre or 12 mile journey. When meeting a group en route it is often advisable to meet at lunch time as this may help to avoid delaying either the group or the Assessor. Patience is the hallmark of the good Assessor. Groups will get lost. This is part of the learning process. If the group manages to sort itself out and reach its destination, it should not present a problem other than that the group will be late, will have covered extra distance and be even more tired and weary than usual.

One of the conditions in the *Award Handbook* is that groups should 'prepare one substantial meal each day'. The Assessor should ensure that groups are adequately nourished, though diet is very much a matter for the individual and there should be a liberal interpretation of what is a substantial meal. Groups may need some encouragement to prepare food at the end of an exhausting day.

Meeting at the End of the Journey – The Debriefing

Assessors should remember that the arrival by the group at their destination is the climax of the journey and the completion of a challenge. Assessors must be aware of this and treat it accordingly. Under no circumstances should the meeting at the end of the venture be turned into an anti-climax and the venture be allowed to end on a low note. If the group is exhaused, they will all appreciate their efforts after a good nights' sleep.

The meeting with the group immediately at the end of the venture is second only in importance to **The First Meeting**. The **Debriefing** is the opportunity for Assessors to congratulate the group on the successful completion of their journey and share in their achievements and the hardships they have overcome. It is concerned with the participants reaction to their venture and the process is helped along by sensitive questioning and prompting on the part of the Assessor and should involve all the participants. The appropriate page of the participants' *Record Books* should be completed and signed after the Debriefing.

Oral presentations, submitted to the Assessor at the end of a venture, **are concerned with the purpose of the venture** and must be considered separately from the debriefing. This distinction should be made clear to the participants. The spontaneous debrief should take place first and followed, after an interval, by the pre-planned oral report. On completion of a satisfactory oral presentation, the Assessor should complete and sign the second appropriate page of the *Record Books*.

If a written, photographic or any other form of presentation is being submitted to the Assessor, or the presentation is being submitted to another approved person, the arrangements should be carefully checked and participants reminded that the failure to submit a report is the most common reason for failing to complete the conditions of the Expeditions Section.

In addition to the prime function of ensuring that the requirements and conditions of the Award are fulfilled, the Assessor has two additional main functions: advising on safety and safeguarding the interests of the Award Scheme.

Advising on Safety

The safety of the participants is the responsibility of the Supervisor, who is the agent of the Operating Authority which has the legal responsibility for the participants' well-being. Although the ultimate responsibility rests with the Supervisor, recommendations which Assessors may make are frequently vital to the safety of the group. The Assessor's local knowledge of the area, weather conditions and hazards such as shooting ranges and mine shafts, make a vital contribution to the participants' safety, as does their detached view of competence and equipment. Supervisors and groups invariably heed the warning of an Assessor. As a last resort, Assessors can withdraw their services, which will bring the venture to an end as a qualifying venture, or as an Award activity, and turn it into a practice journey.

Safeguarding the Interests of the Award Scheme

The Award Scheme relies on Assessors to maintain its reputation in areas where ventures take place. This is largely achieved by Assessors using their local knowledge to avoid sensitive areas, friction with farmers and landowners or overburdened communities. Ensuring an observance of the Country Code during the venture and maintaining good relations with the farmers who provide the camp sites are all part of this task. Very occasionally the Award Scheme relies on the Assessor to resolve problems which come to light after the venture has ended.

GENERAL CONSIDERATIONS

The assessment process outlined above is basically the same for all forms of venture, whatever mode of travel. Assessors may need specialised skills associated with the mode of travel, with ventures abroad, Explorations or Other Adventurous Projects. These specialised skills are considered in the appropriate chapters of this *Guide*, as are greater details on specific aspects such as reports.

SUMMARY
Before the Venture

- Check the *Expedition Notification Forms* to ensure that all the basic conditions and requirements are fulfilled.
- Scrutinise the plans, routes, foul weather alternatives and escape routes as soon as possible to enable the group to make any essential changes.

- Contact the Supervisor (**The Initial Contact**) and arrange a meeting (**The First Meeting**) with the participants and the Supervisor immediately before the venture.
- Find out if a **Local Pre-expedition Check** is to be carried out. If the venture is an Exploration check if a Mentor is involved.
- Ensure that training has been certified in the *Record Books* and that the required number of practice journeys have been completed.
- Check that the group is properly equipped and trained. This task is usually eased if a **Local Pre-expedition Check** has already been carried out.
- Check the arrangements for communication between the group, Supervisor and Assessor. Co-ordinate visits with the Supervisor and pinpoint the location of camp sites and checkpoints.
- Confirm the form and nature of the report and who will receive the submission.

During the Venture

- Meet the group, normally once a day.
- Visit at least one camp site to check camp craft and catering and inspect a camp site after the group has left.
- Check the progress of observations, recording or investigative work associated with the purpose of the venture.
- Assessors, in full consultation with the Supervisor and the group, should not hesitate to require the group to modify the proposed routes for reasons of safety, weather conditions, or a greater understanding of the limited capabilities of group members.

After the Venture

- Meet the group at the end of the journey and carry out an oral **Debriefing** concerned with the participants' reaction and response to their venture. Complete and sign the appropriate page of the *Record Books*.
- **Either** receive an oral presentation, if prior arrangements have been made, relating to the purpose of the venture. This must be distinct and separate from the oral **Debriefing**. Both pages of the *Record Books* may then be completed and signed.
- **Or** confirm the arrangements which have been made concerning the submission of presentations. If the participants have decided to submit their presentations to the Assessor, all concerned should agree when and how this is to happen. On receipt of the presentation, the Assessor signs the other part of the *Record Books* and returns them to the participants.

The role of the Assessor is that of a person who verifies and confirms that the requirements and conditions of the Expeditions Section have been fulfilled rather than one who tests. **There is no pass, no fail, no testing, no marking - either the conditions have been fulfilled or they have not.** There are no written papers or questions as they would not be appropriate for the Expeditions Section. Assessors must have the ability to determine competence by oral questioning and the suitability of equipment by observation. The prime condition is that the participants should carry out a venture of the required distance or duration, in the allotted number of days, by their own physical effort, without outside help or motorised assistance and subsequently, provide a report of their journey.

Joe Cornish

The Assessor Accreditation Scheme

The Duke of Edinburgh's Award, in conjunction with its Wild Country Panels and Operating Authorities, provides an Accreditation Scheme for Wild Country Assessors in order to:

- Ensure a common understanding of the philosophy, spirit and underlying aims of the Award Scheme.
- Ensure a more consistent interpretation of the conditions of the Expeditions Section.
- Establish a positive and supportive approach to assessment.

Wild Country Assessors applying for accreditation fall into two categories:

1. Those who are already members or wish to become members of Wild Country Expedition Panels.

2. Those working through Operating Authorities such as Voluntary Youth Organisations, Further Education Colleges, Local Authorities and industry.

Accredited Assessors will be expected to make a commitment which involves carrying out regular wild country assessments and maintaining a sound working knowledge of the area. This commitment may be supplemented by undertaking Local Pre-expedition Checks or acting as Supervisor or Instructor.

Applicants wishing to become Accredited Assessors must be experienced hill or mountain walkers with an extensive background knowledge of one or more of the designated Wild Country Areas. The Accreditation Scheme is essentially concerned with training Assessors in the aims, requirements and conditions of the Expeditions Section of the Award, and then accrediting them to particular Wild Country Areas. The Accreditation Scheme is not concerned with training in mountain skills or familiarising people with Wild Country Areas.

It is vital that all prospective Assessors are able to relate to young people quickly and easily. The Award Scheme is seeking people who can appreciate the difficulties which many young people face in what is, for many, the greatest challenge they have so far encountered in an environment which they frequently regard as alien.

In the interests of the Award Scheme and Operating Authorities, all applicants should be prepared to permit any necessary checks that may be made by the police under The Protection of Children Act. The Award Scheme, together with other National Voluntary Youth Organisations, does not have direct access, under current legislation, to police screening. However, Local Authorities are required to submit any adult who has regular and substantial contact with young people to such screening by the police.

Applicants should normally be over 21 years of age and preferably over 25 (the maximum age for participants in the Award Scheme) so that the necessary mountain experience and local knowledge, so essential for this role, has been acquired. There is no upper age limit providing that applicants are still practising hill or mountain walkers. The Award welcomes existing Gold Award Holders who have extended their mountain experience or have acquired additional qualifications such as the Mountain Leader Award.

On enrolling onto the Scheme, prospective Accredited Assessors will be entered on a national database, given a number, receive a *Log Book* and then commence the accreditation process.

Recent experience on Assessor Accreditation Courses has revealed that, now that the majority of the experienced Assessors associated with the Wild Country Panels and Operating Authorities have been accredited, many now presenting themselves for accreditation lack the background experience of the Award Scheme and the regular involvement with the Expeditions Section which is so essential in carrying out the role of Assessor effectively.

Prospective Accredited Assessors will be expected to become familiar with the Award, its philosophy and with the conditions of the Expeditions Section through the Award's literature, **before** embarking on the next stage of the accreditation process. This will be followed by attendance at a weekend Accreditation Course and then by an induction period consisting of a number of assessments under the guidance of different experienced Assessors in the Wild Country Area of their choice.

Valuable experience is gained through involvement in supervising practice and qualifying ventures, assessing at Bronze and Silver levels and 'shadowing' and supporting existing Accredited Assessors. Working as an Instructor, or assisting in training always provides a 'grass roots' insight into the Expeditions Section.

The Role of the Assessor

An Accredited Assessor's role will be to:

- Ensure that all participants fulfil the requirements and conditions of the Expeditions Section when undertaking their qualifying venture.
- Use their local knowledge to advise the participants and their Supervisor on safety should this be necessary, though the ultimate responsibility for safety always remains with the Supervisor.
- Safeguard the general interests of The Duke of Edinburgh's Award in the area of the venture.

Additional Functions

In addition, it is expected that all Assessors should be capable of carrying out Local Pre-expedition Checks and, if they wish, be able to assist with Open Golds (see Chapter 2.6).

Local Pre-expedition Checks

A Local Pre-expedition Check is an evaluation of the group's equipment and competence.

This should be carried out in the locality of the group's home base about a week before departing for the qualifying venture in a Wild Country Area. The purpose is to avoid a group travelling, often hundreds of miles, to Wild Country Areas only to find that their equipment or training is unsatisfactory, with all the consequent waste of time, effort, money and distress. If a Local Pre-expedition Check is not carried out, then a similar evaluation of equipment and competence must take place at the **First Meeting** during the acclimatisation period immediately prior to the venture.

Open Golds

An Open Gold is an opportunity for independent participants to join with others who are unable to form a group to carry out their Gold qualifying venture. These last for about a week and are usually organised by Wild Country Panels through the appropriate Territorial or Regional Officer, or by Operating Authorities. The first two or three days are spent in preparation and getting to know each other, and the final four days are devoted to the qualifying venture (see Chapter 2.6 - Open Golds).

THE PRINCIPAL STAGES OF ACCREDITATION
Stage 1 - Wild Country Experience

Persons wishing to become Accredited Assessors who are confident that they have the necessary hill and mountain walking experience and a familiarity with one of the designated Wild Country Areas, should obtain an enrolment form. These are available from Award Head Office and the Award Secretaries and Regional Offices. (See Appendix for the addresses).

Endorsement of their application concerning their hill and mountain walking competence should be sought from a Wild Country Panel Secretary or Operating Authority. the enrolment form shoould then be sent to Award Head Office in Windsor together with the appropriate enrolment fee. The Assessor will then be entered on the national database and issued with an accreditation number and the *Assessor's Log Book*. On receipt of the *Log Book*, the contents should be studied and details of previous experience should be entered on the appropriate pages.

Stage 2 - Introductory Learning and Preparation

Applicants must familiarise themselves with the aims, requirements and conditions through the Award's literature - the *Award Handbook*, *Expedition Guide* and the *Assessor's Log Book*.

447

Practical experience of the Expeditions Section should be expanded by involvement with the supervision of practice and qualifying ventures, assessment of Bronze and Silver Expeditions and instruction, if the prospective Accredited Assessor is not already involved in this work.

This introducory period also provides an excellent opportunity for candidates to acquire qualifications such as the CCPR's Basic Expedition Leaders Award (BELA) and the Mountain Leader Training Board's Award if they do not already hold any national qualifications.

Stage 3 - Accreditation Weekend

When the introductory period has been completed, applicants must attend and satisfactorily complete one of the approved weekend Assessor Accreditation Courses. Around five or six of these courses take place each year in the various Regions of the of the Award Scheme. Details of these courses are advertised in *Award Journal* or are available from the Award's Territorial and Regional Offices.

Stage 4 - Induction Assessments

Applicants must complete a number of Induction Assessments under the guidance of experienced Accredited Assessors in the Wild Country Area of their choice. Some very experienced Assessors who perform a considerable number of assessments may be accredited to more than one designated Wild Country Area.

On satisfactory completion of these stages, the Assessor will become an Accredited Assessor and be issued with an identity card which is initially valid for a period of five years.

OTHER MODES OF TRAVEL

Existing Assessors concerned with water, cycling or horse riding ventures or wishing to assess these modes of travel, should not hesitate to join the Accreditation Scheme. The assessment process is the same regardless of the mode of travel. The Award Scheme will welcome their specialised skills. When completing their enrolment form applicants should make it clear if they wish to be involved with a specific mode of travel.

At Gold level, cycling and horse riding ventures usually take place in designated Wild Country Areas so the majority of the competencies involved are the same as for ventures on foot, in addition to the specialised skills relating to the mode of travel. The Induction Assessments will relate to the mode of travel and they should follow the same procedure as for foot ventures.

For water based ventures at Gold level it is only possible, at the present time, to be accredited to the Severn and Wye Panel or to one of the Wild Country Panels, many of which assess water ventures. The Award Scheme welcomes those with specialist qualifications or skills which can be used in the assessment of water ventures, either with a Wild Country Panel or an Operating Authority. Prospective Assessors should enter their previous water-related experience, any relevant qualifications, familiarise themselves with the philosophy, requirements and conditions through the Award Scheme's literature. They should state whether they wish to be concerned with rowing, canoeing or sailing. Involvement with the Expeditions Section should be extended, if not already involved with this work, and followed by attendance an Accreditation Course.

EXPEDITION GUIDE

Paul Taylor / Oxford Scientific Films

Part

7

APPENDICES

GLOSSARY

WILD COUNTRY AREAS

INDEX

Part 7
Appendices

Appendices

GLOSSARY

Assessor: A person whose principal role is to ensure that the requirements and conditions are fulfilled during qualifying ventures.

Assessor Accreditation Scheme: An 'induction' scheme for those who assess qualifying ventures in Wild Country Areas.

Award Leaders: An adult with the responsibility for co-ordinating Award work with participants and their Adult Helpers.

Award Unit: Units which operate the Scheme for young people on behalf of the Operating Authority.

Conditions: The demands which have to be fulfilled during qualifying ventures specifically associated with one of the three levels of the Award - Bronze, Silver or Gold. For example: the minimum mandatory distance for a Silver Expedition on foot is 48 km.

Expedition: A venture where journeying is the principle component.

Expedition Season: The Expedition season is between the end of March and the end of October. No specific dates are prescribed but the period coincides approximately with British Summer Time.

Exploration: A venture which involves less journeying time and a greater proportion of time to be spent on first-hand investigations or specified activities.

Familiarisation Period: Participants are advised to spend 48 hours in the Wild Country Area to adapt to the environment, adjust to the routine of camp life and prepare themselves and their equipment prior to their departure on the qualifying venture. The First Meeting with the Assessor usually takes place during this period.

Group: A team of between 4 and 7 young people who join together to train for and complete a qualifying venture.

Instructor: A person used by the Unit Leader to provide instruction in one or more specialist aspects of Expedition training.

Local Pre-expedition Check (LPC): A local check of training and equipment carried out by an Assessor, or other competent person, about a week before a group departs for the venture area. It is intended to prevent participants travelling hundreds of miles only to find that their training or equipment is inadequate.

Mode of Travel: All ventures must be carried out by the participants' own physical effort without motorised assistance. The journeys may take place on foot, by cycle, by horse riding or on water by canoeing, sailing or rowing.

Open Award Centre: A centre which provide facilities for unattached and independent participants to take part in the Award Scheme. They are provided by Operating Authorities and extend access to the Award Scheme beyond the confines of schools, youth centres etc.

Open Golds: These are opportunities for participants who are unable to form a group to carry out their Gold venture. They are usually organised by Wild Country Panels or Operating Authorities and last about a week which provides time for a familiarisation and planning period prior to the qualifying venture. All necessary training and practice journeys must have been completed before participants can take part.

Operating Authority: An organisation or body which has been franchised by Award Headquarters to operate the Award Scheme. They range in size from national organisations and Local Government Authorities to independent schools and industrial concerns.

Other Adventurous Projects (OAP): A venture at Gold level of a demanding nature which, though based on a journey, may depart from certain specified conditions of the Award.

Participants: Young people between the ages of 14 and 25 involved in The Duke of Edinburgh's Award .

Practice Journeys: Practice journeys must be completed prior to the qualifying venture. The final practice journey should resemble the qualifying venture as closely as possible in terms of distance each day and terrain.

Presentations: A presentation must be submitted after all ventures. It is the responsibility of the participants to decide on the form and nature of the presentation and to whom it will be submitted.

Purpose: All ventures must have a clearly defined and pre-conceived purpose which provides a focus for the venture .

Qualifying Venture: A venture at Bronze, Silver or Gold level which conforms to the requirements and conditions of the Expeditions Section and which is assessed .

Requirements: The demands which are common to all ventures regardless of the level of the Award and which have to be satisfied in a qualifying venture. E.g. accommodation must be by camping.

Responsible Person: A person, usually associated with the Operating Authority or Award Unit, who is familiar with the location and timing of the venture and who may be contacted in the event of an emergency.

Supervisor: A person who is responsible to the Operating Authority for the safety and well-being of the participants during the qualifying venture.

Wild Country Area: One of a number of designated areas of country where Gold level land-based ventures should take place.

Wild Country Panels: A team of Assessors associated with each of the designated Wild Country Areas.

COUNTRY CODE

- Enjoy the countryside and respect its life and work.
- Guard against all risk of fire.
- Fasten all gates.
- Keep your dogs under close control.
- Keep to rights-of-way across farmland.
- Use gates and stiles to cross fences, hedges and walls.
- Leave livestock, crops and machinery alone.
- Take your litter home.
- Help to keep all water clean.
- Protect wildlife, plants and trees.
- Take special care on country roads.
- Make no unnecessary noise.

MOUNTAIN BIKE CODE

Rights-of-Way:
- Bridleways - open to cyclists, but you must give way to walkers and horseriders.
- Byways - usually unsurfaced tracks open to cyclists. As well as walkers and horseriders, you may meet occasional vehicles which also have right of access
- Public footpaths - no right to cycle exists. Look out for finger posts from the highway or waymarking arrows - blue for bridleways, red for byways, yellow for footpaths.
- Please note - these rights-of-way do not apply in Scotland.

Other Access:
- Open land - on most upland, moorland and farmland cyclists normally have no right of access without express permission from the landowner.
- Towpaths - a British Waterways permit is required for cyclists wishing to use their canal towpaths.
- Pavements - cycling is not permitted on pavements.
- Designated cycle paths - look out for designated cycle paths or bicycle routes which may be found in urban areas, on Forestry Commission land, on disused railway lines and other open spaces.

General Information:
- Cyclists must adhere to the Highway Code.

Safety:
- Ensure that your bike is safe to ride and be prepared for all emergencies.
- You are required by law to carry working lights after dark.
- Always carry some form of identification.
- Always tell someone where you are going.
- Learn to apply the basic principles of first aid.
- Reflective materials on your clothes or bike can save your life.
- For safety on mountains refer to the British Mountaineering Council's publication 'Safety on Mountains'.
- Ride under control downhill since this is when serious accidents often occur.
- If you intend to ride fast off-road, it is advisable to wear a helmet.
- Particular care should be taken on unstable or wet surfaces.

WATER SPORTS CODE

Respect the feelings of the local community:
- Keep to authorised routes, footpaths, access points and slipways.
- Leave gates as you found them.
- Do not obstruct farm entrances, footpaths, access points or narrow lanes with parked vehicles.
- Do not leave litter.
- Do not shout or make a lot of noise.

Safety:
- Know how to control your craft properly.
- Make sure your craft is in good condition and properly equipped.
- Be familiar with the warning signs and markers for other activities.
- Keep a good look out for people in the water, slow moving boats and anglers concealed in bankside vegetation.
- Respect other water users and be willing to give way or stop.
- Have and display all appropriate licences.
- Launch and moor only at authorised sites.
- Make sure you are adequately insured.
- Do not cut across the bows of an oncoming craft.
- Do not waste water by passing a single craft through a lock.
- Do not take your craft onto private waters without permission.
- Do not launch, moor or land on private property without permission.
- Do not use private waters to gain access to other waters or property.
- Do not allow fuel, oil, paint, chemicals or detergents to leak or spill into the water or onto the bank.

Simple Guidelines for River Canoeists:

- Unless otherwise indicated, presume that there is no right of navigation along a river and make arrangements as required before gaining access. If in doubt contact the BCU or your Regional Office of the National Rivers Authority.
- Access to many rivers is dependent upon the goodwill of riparian/fishery owners and tenants. Do not abuse their goodwill.
- Respect local rules enforced by the BCU and affiliated canoeing clubs.
- Do not enter private land without permission.
- Have respect for the interests of nature conservation, anglers, agriculture and any other users of the river.
- Stop canoeing if you are clearly disturbing wildlife.
- Pass anglers quietly. Try to:
  - keep away from banks being fished.
  - comply with reasonable directional requests.
  - keep well clear of fishing tackle.
  - avoid loitering in pools if anyone is fishing.
  - cause as little disturbance as possible.
- Leave the water if requested to do so by an NRA bailiff. Ask for identification if you are in doubt.

Canoeing is a safe sport but:

- Take all reasonable safety precautions.
- Wear a buoyancy aid.
- Keep away from weirs, sluices and other dangers.
- Do not venture into flood-swollen rivers.
- Cover cuts and scratches.
- Avoid swallowing river water.

BIBLIOGRAPHY

Award Handbook

The Duke of Edinburgh's Award
ISBN 0-905425-10-3

Exploration Resource Pack

The Duke of Edinburgh's Award

**Land Navigation - Route
Finding with Map and Compass**

Wally Keay
The Duke of Edinburgh's Award
ISBN 0-905425-06-5

Playback
A Guide to Reviewing Activities

Roger Greenaway,
The Duke of Edinburgh's Award
ISBN 0-905425-09-X

**Workshop - Design and Make
Expedition Equipment**

Don Robertson
The Duke of Edinburgh's Award
ISBN 0-905425-07-3

****

Cycling Off-Road and the Law

Neil Horton
Cyclists Touring Club

Directory of Tests and Awards

British Canoe Union Coaching
Service

Effective Leadership

John Adair
Pan Business Management
ISBN 0-330-28100-3

First Aid Manual
The Combined Manual

St. John Ambulance
St. Andrew's Ambulance Association
British Red Cross Society
ISBN 0-86318-978-4.

Follow the Map
An Ordnance Survey Guide

John G. Wilson
ISBN 0-7136-2459-0

Mountain Bikes

Usborne Superskills
ISBN 0-7460-0520-2

459

Mountaincraft and Leadership
Eric Langmuir
Mountain Leader Training Board
ISBN 1-85060-295-6

RYA Coaching Manual
The Royal Yachting Association

Safety on Mountains
British Mountaineering Council.
ISBN 0-903908-95-6

Sailing - Know the Game
Adlard Coles
ISBN 0-7136360-78-5

The Ultimate Guide for Fitness
Chantal Gosselin
Reebok
ISBN 0-09-178370-4

Tread Lightly
British Mountaineering Council

Weather at Sea
Houghton

Weather for Hillwalkers and Climbers - a guide to keeping out of trouble
Malcolm Thomas
Alan Sutton Publishing Ltd.

USEFUL ADDRESSES

Adventure Activities Licensing Authority: 17 Lambourne Crescent, Llanishen, Cardiff CF4 5GG. Tel: 01222 755715.

Amateur Rowing Association: The Priory, 6 Lower Mall, London W6 9DJ. Tel: 020 8741 5314.

British Canoe Union: John Dudderidge House, Adbolton Lane, West Bridgford, Nottingham NG2 5AS. Tel: 01159 821100.

British Cycling Federation: The National Cycling Centre, Stuart Street, Manchester M11 4DQ. Tel: 0161 230 2301.

British Horse Society: British Equestrian Centre, Stoneleigh Park, Kenilworth, Warwickshire CV8 2LR. Tel: 01203 696697.

British Mountaineering Council: 177-179 Burton Road, West Didsbury, Manchester M20 2BB. Tel: 0161 445 4747.

British Red Cross Society: 9 Grosvenor Crescent, London SW1X 7EJ. Tel: 020 7235 5454.

British Sports Trust (for BELA Award): Francis House, Francis Street, London SW1P 1DE. Tel: 020 7828 3163.

British Waterways Board: Willow Grange, Church Road Watford, Hertfordshire WD1 3QA. Tel: 01923 226422.

Commonwealth Youth Exchange Council: 7 Lion Yard, Tremadoc Road, Clapham, London SW4 7NQ. Tel: 020 7498 6151.

Cyclists Touring Club: Cottrell House, 69 Meadrow, Godalming, Surrey GU7 3HS. Tel: 01483 417217.

Expedition Advisory Centre: Royal Geographical Society, 1 Kensington Gore, London SW7 2AR. Tel: 020 7591 3030.

Forestry Commission: 231 Corstorphine Road, Edinburgh EH12 7AT. Tel: 0131 334 2576.

Mountain Leader Training Board: Capel Curig, Gwynedd LL24 0ET. Tel: 01690 720314.

National Rivers Authority: Rio House, Waterside Drive, Aztec West, Almondsbury, Bristol BS12 4UD. Tel: 01454 624400.

National Navigational Award: Royal Institute of Navigation, 1 Kensington Gore, London SW7 2AT. Tel: 020 7589 5021.

Outward Bound Trust: Watermillock, Nr. Penrith, Cumbria CA11 0JL. Tel: 0990 134227.

Royal Geographic Society: 1 Kensington Gore, London SW7 2AR. Tel: 020 7589 5466.

Royal Society for the Prevention of Accidents (RoSPA): Edgbaston Park, 353 Bristol Road, Birmingham B5 7ST. Tel: 0121 248 2000.

Royal Yachting Association: RYA House, Romsey Road, Eastleigh, Hampshire SO50 9YA. Tel: 023 8062 7400.

St. Andrew's Ambulance Association: 48 Milton Street, Glasgow G4 0HR. Tel: 0141 332 4031.

St. John Ambulance: 1 Grosvenor Crescent, London SW1X 7EF. Tel: 020 7235 5231.

Young Explorers Trust: c/o Royal Geographic Society: 1 Kensington Gore, London SW7 2AR. Tel: 01623 861027.

Youth Exchange Centre: The British Council, 10 Spring Gardens, London SW1A 2BN. Tel: 020 7389 4030.

Youth Hostels Association: Trevelyan House, 8 St. Stephen's Hill, St. Albans, Hertfordshire AL1 2DY. Tel: 01727 855215.

INDEX

WILD COUNTRY EXPEDITION PANEL AREAS

Wild Country Expedition Panels for The United Kingdom

1. Western Isles (WI)
2. Caithness & Sutherland (CS)
3. Wester Ross (WR)
4. Easter Ross (ER)
5. Skye & Lochalsh (SL)
6. Inverness (IN)
7. Isle of Mull (IM)
8. Lochaber District (LG)
9. Grampian & Cairngorm (GR)
10. Tayside (TY)
11. Trossachs & Crianlarich (TC)
12. Lomond & Argyll (LA)
13. Isle of Arran (IA)
14. Galloway Hills (GH)
15. Lowther Hills (LH)
16. Scottish Borders (SB)
17. Sperrin Mountains (SM)
18. North Antrim Hills (AH)
19. Mourne Mountains (MM)
20. Isle of Man (MN)
21. Cumbria (CU)
22. Cheviots (CH)
23. Durham Dales (DD)
24. North Yorkshire Pennines (YP)
25. Yorkshire Dales (YD)
26. North York Moors (YM)
27. Peak District (PD)
28. Snowdonia (SO)
29. Mid Wales (MW)
30. Brecon Beacons & Black Mountains (BX)
31. Dartmoor (DA)

Also Severn & Wye Panel (SW) and, in Germany: Bavaria Panel (BA)

The Panel Secretaries address list is published annually in the Spring edition of *Award Journal.*

PORTABLE BREATHING APPARATUS

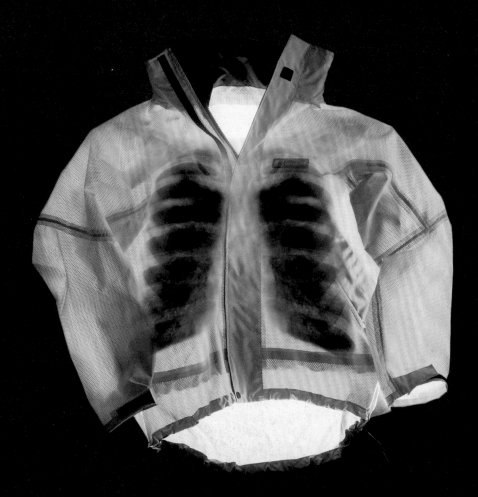

The Berghaus PacLite® jacket. Made from the innovative GORE-TEX®
PacLite® fabric. 100% waterproof. Extremely breathable.
Highly packable. How can you survive without it?

GORE-TEX®, PACLITE® and designs are registered trademarks
of W L Gore and Associates.

For details of your nearest stockists call 0191 516 5600

STRAP IN FOR THE RIDE OF YOUR LIFE

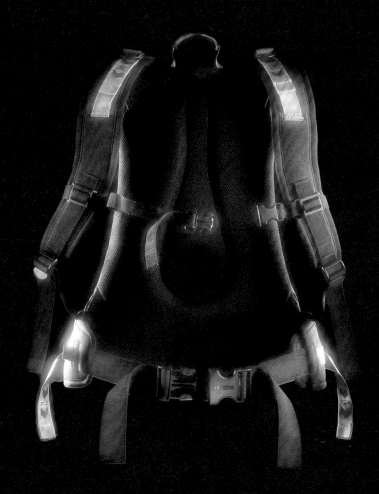

MOUNTAIN BIKING ■ SNOWBOARDING ■ CLIMBING ■ SKIING ■ TRAIL-RUNNING

Whenever you're set to take off, the Berghaus limpet carrying and compression system will guarantee your rucsac sticks with you. Winner of a **Millennium Design Award** for style and innovation, this unique strapping system is an integrated part of the sac and exclusive to the Berghaus endurance rucsac range. Once the straps are adjusted the load is evenly distributed throughout the sac and firmly moulds to your back - leaving you free to have the ride of your life.

www.berghaus.com

THE DUKE OF EDINBURGH'S AWARD

Head Office:

Gulliver House, Madeira Walk, WINDSOR, Berkshire SL4 1EU.
Tel: 01753 727400. Fax: 01753 810666.
E-mail: ops@theaward.org   web: www.theaward.org

The Award Scheme Ltd:

Unit 18/19 Stewartfield Industrial Estate, off Newhaven Road,
EDINBURGH EH6 5RQ. Tel: 0131 553 5280. Fax: 0131 553 5776.
E-mail: asl@theaward.org

UK Award Offices:

- **Northern Ireland:** 28 Wellington Park, BELFAST BT9 6DL
  Tel: 028 9050 9550 Fax: 028 9050 9555
  E-mail: nireland@theaward.org   web: www.theaward.org/northernireland
- **Scotland:** 69 Dublin Street, EDINBURGH EH3 6NS
  Tel: 0131 556 9097 Fax: 0131 557 8044
  E-mail: scotland@theaward.org   web: www.theaward.org/scotland
- **Wales:** Oak House, 12 The Bulwark, BRECON, Powys LD3 7AD
  Tel: 01874 623086 Fax: 01874 611967
  E-mail: wales@theaward.org   web: www.theaward.org/wales
- **East Midlands:** c/o Chilwell Comprehensive School, Queens Road West, Beeston,
  NOTTINGHAM NG9 5AL
  Tel: 0115 922 8002 Fax: 0115 922 8302
  E-mail: eastmid@theaward.org   web: www.theaward.org/eastmidlands
- **West Midlands:** 89-91 Hatchett Street, Newtown, BIRMINGHAM BI9 3NY
  Tel: 0121 359 5900 Fax: 0121 359 2933
  E-mail: westmid@theaward.org   web: www.theaward.org/westmidlands
- **South East:** 10 Station Road, CHERTSEY, Surrey KT16 8BE
  Tel: 01932 564800 Fax: 01932 564788
  E-mail: southeast@theaward.org   web: www.theaward.org/southeast
- **South West:** Court Gatehouse, Corsham Court, CORSHAM, Wiltshire SN13 0BZ
  Tel: 01249 701000 Fax: 01249 701050
  E-mail: southwest@theaward.org   web: www.theaward.org/southwest
- **North East:** Maritime Chambers, 1 Howard Street, NORTH SHIELDS,
  Tyne & Wear NE30 1LZ
  Tel: 0191 270 3000 Fax: 0191 270 3007
  E-mail: northeast@theaward.org   web: www.theaward.org/northeast
- **North West:** Churchgate House, 56 Oxford Street, MANCHESTER M1 6EU
  Tel: 0161 228 3688 Fax: 0161 228 3960
  E-mail: northwest@theaward.org   web: www.theaward.org/northwest
- **East:** 17 Lower Southend Road, WICKFORD, Essex SS11 8ES
  Tel: 01268 571393 Fax: 01268 562060
  E-mail: east@theaward.org   web: www.theaward.org/east
- **London:** 7th Floor, Therese House, 29-30 Glasshouse Yard, LONDON EC1A 4JN
  Tel: 020 7253 5544 Fax: 020 7253 5224
  E-mail: london@theaward.org   web: www.theaward.org/london

International Award Association:

Award House, 7-11 St. Matthew Street, LONDON SW1P 2JT.
Tel: 020 7222 4242. Fax: 020 7222 4141.
E-mail: sect@intaward.org Web: www.intaward.org.